Chemistry Study Guide

Gregory Howard Gebhart

Copyright © 2017 Gregory Howard Gebhart

Chemistry Study Guide
All Rights Reserved

ISBN-13: 978-1977536105

ISBN-10: 1977536107

Dedication

This book is dedicated to my wife, Fran, my mother, Eve, my father, Howard, my brothers, Brian & Eric, and my parents-in-law, Gary & Doris. It is also dedicated to my teachers and professors and to my supervisors and co-workers. Finally, it is dedicated to the one, true, living God.

Chapter 1
A First Look at Chemistry

According to the Merriam-Webster Dictionary, "Chemistry is the study of matter." Everything on Earth and in the surrounding universe is made of matter. Chemistry studies every aspect of this matter. Therefore Chemistry is the study of every aspect of everything of which the physical universe is made.

Chemistry studies the physical composition of matter: the elements and chemicals of which all matter is made. In turn, Chemistry studies the make-up of the elements and chemicals – the structure of their individual atoms (protons, neutrons and electrons) and the chemical bonds between the atoms. It studies the different physical states of the elements and chemicals that make up all matter (gaseous, liquid and solid states). For instance, in their gaseous state, Chemistry studies the effects of pressure on the volume of gaseous elements and chemicals and visa versa. For example, in their liquid state Chemistry studies mixtures and solutions of chemicals and the acid and bases behaviors of chemicals in water (aqueous) solutions.

It studies the exact products that the chemicals in the physical universe can produce in chemical reactions, whether they are molecules, ionic compounds, or pure elements. When studying how chemicals react, Chemistry determines not only what products are produced, but also what amounts of chemical reactants form what amounts of chemical products, the heat produced or absorbed by the reactions, and other changes in the physical states of the reactants and products. Chemistry studies everything from pure elements like gold or sulfur found in nature to very large biochemicals found in living organisms.

We will begin learning about Chemistry by starting with what is now known about matter, about the elements that make it up. Then we will go back in history to learn how humanity discovered all of these things about matter over the centuries. We will review not only the successes, but also the misconceptions and failures of the study of Chemistry. This will be a historical and conceptual study of Chemistry.

When you come into a Chemistry classroom one of the first things you usually see is a large periodic table of the elements. This is because the periodic table summarizes much of what is known about the elements that make up all matter in the universe. If you learn to understand the periodic table, even without knowing anything about an element to start with, you can tell an awful lot about it by knowing about its position on the table and the adjacent elements on the table. Let us take a look at several different versions of the periodic table of the elements.

We will go over them in much greater detail later, but right now we want you to see how simple and yet elegant the periodic table of the elements is, how easily you can learn so much from it. Don't try to memorize it, just look over briefly each version as you come to it. There will be more time later to go into greater detail.

Here is a simplified version of the Periodic Table of the Elements showing only the different types of atoms (the different elements) and their symbols (which are either just one letter or two):

H Hydrogen																	He Helium
Li Lithium	Be Beryllium											B Boron	C Carbon	N Nitrogen	O Oxygen	F Fluorine	Ne Neon
Na Sodium	Mg Magnesium											Al Aluminum	Si Silicon	P Phosphorus	S Sulfur	Cl Chlorine	Ar Argon
K Potassium 39.10	Ca Calcium 40.08	Sc Scandium 44.96	Ti Titanium	V Vanadium	Cr Chromium	Mn Manganese	Fe Iron	Co Cobalt	Ni Nickel	Cu Copper	Zn Zinc	Ga Gallium	Ge Germanium	As Arsenic	Se Selenium	Br Bromine	Kr Krypton
Rb Rubidium 85.469	Sr Strontium 87.82	Y Yttrium 88.908	Zr Zirconium	Nb Nobelium	Mo Molybdenum	Tc Technetium	Ru Ruthenium	Rh Rhodium	Pd Palladium	Ag Silver	Cd Cadmium	In Indium	Sn Tin	Sb Antimony	Te Tellurium	I Iodine	Xe Xenon
Cs Cesium	Ba Barium	Lu Lutetium	Hf Hafnium	Ta Tantalum	W Tungsten	Re Rhenium	Os Osmium	Ir Iridium	Pt Platinum	Au Gold	Hg Mercury	Tl Thallium	Pb Lead	Bi Bismuth	Po Polonium	At Astatine	Rn Radon
Fr Francium	Ra Radium	Lr Lawrencium	Unq Unnilquadium	Unp Unnilpentium	Unh Unnilhexium	Uns Unnilseptium	Uno Unniloctium	Uun Unniunium	Uuu Unununium	Uub							

		La Lanthanum	Ce Cesium	Pr Praseodymium	Nd Neodymium	Pm Promethium	Sm Samarium	Eu Europium	Gd Gadolinium	Tb Terebium	Dy Dysprosium	Ho Holmium	Er Erbium 167.26	Tm Thulium	Yb Ytterbium
		Ac Actinium	Th Thallium	Pa Protactinium	U Uranium	Np Neptunium	Pu Plutonium	Am Americium	Cm Curium	Bk Berkelium	Cf Californium	Es Einsteinium	Fm Fermium	Md Mendelevium	No Nobelium

Here is the same Periodic Table of the Elements that shows the Elements' symbols and names, but with their Atomic Numbers (the whole numbers above their Symbols; all of the atoms of a given elements have the same Atomic Number and that Atomic Number is the number of protons in the nucleus – we'll learn more about this later).

1 H Hydrogen																	2 He Helium
3 Li Lithium	4 Be Beryllium											5 B Boron	6 C Carbon	7 N Nitrogen	8 O Oxygen	9 F Fluorine	10 Ne Neon
11 Na Sodium	12 Mg Magnesium											13 Al Aluminum	14 Si Silicon	15 P Phosphorus	16 S Sulfur	17 Cl Chlorine	18 Ar Argon
19 K Potassium	20 Ca Calcium	21 Sc Scandium	22 Ti Titanium	23 V Vanadium	24 Cr Chromium	25 Mn Manganese	26 Fe Iron	27 Co Cobalt	28 Ni Nickel	29 Cu Copper	30 Zn Zinc	31 Ga Gallium	32 Ge Germanium	33 As Arsenic	34 Se Selenium	35 Br Bromine	36 Kr Krypton
37 Rb Rubidium	38 Sr Strontium	39 Y Yttrium	40 Zr Zirconium	41 Nb Nobelium	42 Mo Molybdenum	43 Tc Technetium	44 Ru Ruthenium	45 Rh Rhodium	46 Pd Palladium	47 Ag Silver	48 Cd Cadmium	49 In Indium	50 Sn Tin	51 Sb Antimony	52 Te Tellurium	53 I Iodine	54 Xe Xenon
55 Cs Cesium	56 Ba Barium	71 Lu Lutetium	72 Hf Hafnium	73 Ta Tantalum	74 W Tungsten	75 Re Rhenium	76 Os Osmium	77 Ir Iridium	78 Pt Platinum	79 Au Gold	80 Hg Mercury	81 Tl Thallium	82 Pb Lead	83 Bi Bismuth	84 Po Polonium	85 At Astatine	86 Rn Radon
87 Fr Francium	88 Ra Radium	103 Lr Lawrencium	104 Unq Unnilquadium	105 Unp Unnilpentium	106 Unh Unnilhexium	107 Uns Unnilseptium	108 Uno Unniloctium	109 Uun Unnunilium	110 Uuu Unununium	111 Uub	112						

57 La Lanthanum	58 Ce Cesium	59 Pr Praseodymium	60 Nd Neodymium	61 Pm Promethium	62 Sm Samarium	63 Eu Europium	64 Gd Gadolinium	65 Tb Terebium	66 Dy Dysprosium	67 Ho Holmium	68 Er Erbium	69 Tm Thulium	70 Yb Ytterbium
89 Ac Actinium	90 Th Thallium	91 Pa Protactinium	92 U Uranium	93 Np Neptunium	94 Pu Plutonium	95 Am Americium	96 Cm Curium	97 Bk Berkelium	98 Cf Californium	99 Es Einsteinium	100 Fm Fermium	101 Md Mendelevium	102 No Nobelium

This is pretty much what most Periodic Tables of the Elements look like. You will notice that decimal numbers or numbers in parentheses have been added under the names of the elements. These are the average atomic masses of each element. The periodic table lists the elements in order by their atomic masses starting with the lightest, hydrogen (H) through the heaviest. We will use them extensively later on in the course when we get to

1 H Hydrogen 1.01																	2 He Helium 4.00
3 Li Lithium 6.94	4 Be Beryllium 9.01											5 B Boron 10.81	6 C Carbon 12.01	7 N Nitrogen 14.01	8 O Oxygen 16.00	9 F Fluorine 19.00	10 Ne Neon 20.18
11 Na Sodium 22.99	12 Mg Magnesium 24.31											13 Al Aluminum 26.98	14 Si Silicon 28.09	15 P Phosphorus 30.97	16 S Sulfur 32.07	17 Cl Chlorine 35.45	18 Ar Argon 83.90
19 K Potassium 39.10	20 Ca Calcium 40.08	21 Sc Scandium 44.96	22 Ti Titanium 47.87	23 V Vanadium 50.94	24 Cr Chromium 52.00	25 Mn Manganese 54.94	26 Fe Iron 55.85	27 Co Cobalt 58.93	28 Ni Nickel 58.69	29 Cu Copper 63.54	30 Zn Zinc 65.39	31 Ga Gallium 67.72	32 Ge Germanium 72.61	33 As Arsenic 74.92	34 Se Selenium 78.96	35 Br Bromine 79.90	36 Kr Krypton 83.80
37 Rb Rubidium 85.469	38 Sr Strontium 87.82	39 Y Yttrium 88.908	40 Zr Zirconium 91.22	41 Nb Nobelium 92.906	42 Mo Molybdenum 95.94	43 Tc Technetium (97)	44 Ru Ruthenium 102.91	45 Rh Rhodium 102.91	46 Pd Palladium 106.4	47 Ag Silver 107.87	48 Cd Cadmium 112.4	49 In Indium 114.82	50 Sn Tin 116.69	51 Sb Antimony 121.75	52 Te Tellurium 127.6	53 I Iodine 126.9	54 Xe Xenon 131.3
55 Cs Cesium 132.91	56 Ba Barium 137.33	71 Lu Lutetium 174.97	72 Hf Hafnium 172.49	73 Ta Tantalum 190.95	74 W Tungsten 183.85	75 Re Rhenium 186.21	76 Os Osmium 190.2	77 Ir Iridium 192.22	78 Pt Platinum 195.09	79 Au Gold 196.97	80 Hg Mercury 200.59	81 Tl Thallium 204.37	82 Pb Lead 207.2	83 Bi Bismuth 208.96	84 Po Polonium (208)	85 At Astatine (210)	86 Rn Radon (222)
87 Fr Francium (223)	88 Ra Radium 226.03	103 Lr Lawrencium (260)	104 Unq Unnilquadium (261)	105 Unp Unnilpentium (262)	106 Unh Unnilhexium (263)	107 Uns Unnilseptium (262)	108 Uno Unniloctium (265)	109 Uun Ununilium (269)	110 Uun Ununilium (269)	111 Uuu Unununium (272)	112 Uub						

57 La Lanthanum 138.91	58 Ce Cesium 140.12	59 Pr Praseodymium 140.907	60 Nd Neodymium 144.24	61 Pm Promethium (145)	62 Sm Samarium 150.36	63 Eu Europium 151.965	64 Gd Gadolinium 157.25	65 Tb Terebium 158.93	66 Dy Dysprosium 162.50	67 Ho Holmium 164.93	68 Er Erbium 167.26	69 Tm Thulium 168.93	70 Yb Ytterbium 173.04
89 Ac Actinium (227)	90 Th Thallium 232.04	91 Pa Protactinium 231.03	92 U Uranium 238.029	93 Np Neptunium (237)	94 Pu Plutonium (244)	95 Am Americium (243)	96 Cm Curium (247)	97 Bk Berkelium (247)	98 Cf Californium (251)	99 Es Einsteinium (254)	100 Fm Fermium (257)	101 Md Mendelevium (258)	102 No Nobelium (259)

In this version of the Periodic Table of the Elements, you are introduced to the states of matter that each pure element is found in at what chemists call Standard Temperature and Pressure (STP). The elements that are solid at STP are shown in black characters, the elements that are liquid at STP are shown in blue characters, and the elements that are gases at STP are shown in red characters.

1 H Hydrogen 1.01																	2 He Helium 4.00
3 Li Lithium 6.94	4 Be Beryllium 9.01											5 B Boron 10.81	6 C Carbon 12.01	7 N Nitrogen 14.01	8 O Oxygen 16.00	9 F Fluorine 19.00	10 Ne Neon 20.18
11 Na Sodium 22.99	12 Mg Magnesium 24.31											13 Al Aluminum 26.98	14 Si Silicon 28.09	15 P Phosphorus 30.97	16 S Sulfur 32.07	17 Cl Chlorine 35.45	18 Ar Argon 83.90
19 K Potassium 39.10	20 Ca Calcium 40.08	21 Sc Scandium 44.96	22 Ti Titanium 47.87	23 V Vanadium 50.94	24 Cr Chromium 52.00	25 Mn Manganese 54.94	26 Fe Iron 55.85	27 Co Cobalt 58.93	28 Ni Nickel 58.69	29 Cu Copper 63.54	30 Zn Zinc 65.39	31 Ga Gallium 67.72	32 Ge Germanium 72.61	33 As Arsenic 74.92	34 Se Selenium 78.96	35 Br Bromine 79.90	36 Kr Krypton 83.80
37 Rb Rubidium 85.469	38 Sr Strontium 87.82	39 Y Yttrium 88.908	40 Zr Zirconium 91.22	41 Nb Nobelium 92.906	42 Mo Molybdenum 95.94	43 Tc Technetium (97)	44 Ru Ruthenium 102.91	45 Rh Rhodium 102.91	46 Pd Palladium '106.4	47 Ag Silver 107.87	48 Cd Cadmium 112.4	49 In Indium 114.82	50 Sn Tin 116.69	51 Sb Antimony 121.75	52 Te Tellurium 127.6	53 I Iodine 126.9	54 Xe Xenon 131.3
55 Cs Cesium 132.91	56 Ba Barium 137.33	57 La Lanthanum 138.91	72 Hf Hafnium 172.49	73 Ta Tantalum 190.95	74 W Tungsten 183.85	75 Re Rhenium 186.21	76 Os Osmium 190.2	77 Ir Iridium 192.22	78 Pt Platinum 195.09	79 Au Gold 196.97	80 Hg Mercury 200.59	81 Tl Thallium 204.37	82 Pb Lead 207.2	83 Bi Bismuth 208.96	84 Po Polonium (208)	85 At Astatine (210)	86 Rn Radon (222)
87 Fr Francium (223)	88 Ra Radium 226.03	103 Lr Lawrencium (260)	104 Unq Unnilquadium (261)	105 Unp Unnilpentium (262)	106 Unh Unnilhexium (263)	107 Uns Unnilseptium (262)	108 Uno Unniloctium (265)	109 Uun Unnunilium (269)	110 Uun Unnunilium (269)	111 Uuu Unununium (272)	112 Uub						

	71 Lu Lutetium 174.97	58 Ce Cesium 140.12	59 Pr Praseodymium 140.907	60 Nd Neodymium 144.24	61 Pm Promethium (145)	62 Sm Samarium 150.36	63 Eu Europium 151.965	64 Gd Gadolinium 157.25	65 Tb Terebium 158.93	66 Dy Dysprosium 162.50	67 Ho Holmium 164.93	68 Er Erbium 167.26	69 Tm Thulium 168.93	70 Yb Ytterbium 173.04
	89 Ac Actinium (227)	90 Th Thallium 232.04	91 Pa Protactinium 231.03	92 U Uranium 238.029	93 Np Neptunium (237)	94 Pu Plutonium (244)	95 Am Americium (243)	96 Cm Curium (247)	97 Bk Berkelium (247)	98 Cf Californium (251)	99 Es Einsteinium (254)	100 Fm Fermium (257)	101 Md Mendelevium (258)	102 No Nobelium (259)

Periodic Table of the Elements

In this version of the Periodic Table of the Elements, you are introduced to the different classifications of the pure elements. The elements that are metals have kept their boxes' black borders. The elements that are semimetals are enclosed with blue borders and the elements that are nonmetals are enclosed with red borders.

1 H Hydrogen 1.01																	2 He Helium 4.00
3 Li Lithium 6.94	4 Be Beryllium 9.01											5 B Boron 10.81	6 C Carbon 12.01	7 N Nitrogen 14.01	8 O Oxygen 16.00	9 F Fluorine 19.00	10 Ne Neon 20.18
11 Na Sodium 22.99	12 Mg Magnesium 24.31											13 Al Aluminum 26.98	14 Si Silicon 28.09	15 P Phosphorus 30.97	16 S Sulfur 32.07	17 Cl Chlorine 35.45	18 Ar Argon 83.90
19 K Potassium 39.10	20 Ca Calcium 40.08	21 Sc Scandium 44.96	22 Ti Titanium 47.87	23 V Vanadium 50.94	24 Cr Chromium 52.00	25 Mn Manganese 54.94	26 Fe Iron 55.85	27 Co Cobalt 58.93	28 Ni Nickel 58.69	29 Cu Copper 63.54	30 Zn Zinc 65.39	31 Ga Gallium 67.72	32 Ge Germanium 72.61	33 As Arsenic 74.92	34 Se Selenium 78.96	35 Br Bromine 79.90	36 Kr Krypton 83.80
37 Rb Rubidium 85.469	38 Sr Strontium 87.82	39 Y Yttrium 88.908	40 Zr Zirconium 91.22	41 Nb Nobelium 92.906	42 Mo Molybdenum 95.94	43 Tc Technetium (97)	44 Ru Ruthenium 102.91	45 Rh Rhodium 102.91	46 Pd Palladium '106.4	47 Ag Silver 107.87	48 Cd Cadmium 112.4	49 In Indium 114.82	50 Sn Tin 116.69	51 Sb Antimony 121.75	52 Te Tellurium 127.6	53 I Iodine 126.9	54 Xe Xenon 131.3
55 Cs Cesium 132.91	56 Ba Barium 137.33	71 Lu Lutetium 174.97	72 Hf Hafnium 172.49	73 Ta Tantalum 190.95	74 W Tungsten 183.85	75 Re Rhenium 186.21	76 Os Osmium 190.2	77 Ir Iridium 192.22	78 Pt Platinum 195.09	79 Au Gold 196.97	80 Hg Mercury 200.59	81 Tl Thallium 204.37	82 Pb Lead 207.2	83 Bi Bismuth 208.96	84 Po Polonium (208)	85 At Astatine (210)	86 Rn Radon (222)
87 Fr Francium (223)	88 Ra Radium 226.03	103 Lr Lawrencium (260)	104 Unq Unnilquadium (261)	105 Unp Unnilpentium (262)	106 Unh Unnilhexium (263)	107 Uns Unnilseptium (262)	108 Uno Unniloctium (265)	110 Uun Ununilium (269)	111 Uuu Unununium (272)	112 Uub							

57 La Lanthanum 138.91	58 Ce Cesium 140.12	59 Pr Praseodymium 140.907	60 Nd Neodymium 144.24	61 Pm Promethium (145)	62 Sm Samarium 150.36	63 Eu Europium 151.965	64 Gd Gadolinium 157.25	65 Tb Terebium 158.93	66 Dy Dysprosium 162.50	67 Ho Holmium 164.93	68 Er Erbium 167.26	69 Tm Thulium 168.93	70 Yb Ytterbium 173.04
89 Ac Actinium (227)	90 Th Thallium 232.04	91 Pa Protactinium 231.03	92 U Uranium 238.029	93 Np Neptunium (237)	94 Pu Plutonium (244)	95 Am Americium (243)	96 Cm Curium (247)	97 Bk Berkelium (247)	98 Cf Californium (251)	99 Es Einsteinium (254)	100 Fm Fermium (257)	101 Md Mendelevium (258)	102 No Nobelium (259)

In this version of the periodic table of the elements you see the different types of the outermost atomic orbitals (s, p, d, f) that the "valence" electrons of the elements are found in. The negatively charged electrons "orbit" their atoms' small nuclei (where the positively charged protons and electrically neutral neutrons are). This gives the s-block, the p-block, the d-block, and the f-block elements.

Periodic Table of the Elements

In this version of the periodic table of the elements you see the different types of the outermost atomic orbitals (s, p, d, f) that the "valence" electrons of the elements are found in. The negatively charged electrons "orbit" their atoms' small nuclei (where the positively charged protons and electrically neutral neutrons are). This gives the s-block, the p-block, the d-block, and the f-block elements.

Group	1 (I)	2 (II)	3	4	5	6	7	8	9	10	11	12	13 (III)	14 (IV)	15 (V)	16 (VI)	17 (VII)	18 (VIII)
1	1 H Hydrogen 1.01 $1s^1$																	2 He Helium 4.00
2	3 Li Lithium 6.94 $2s^1$	4 Be Beryllium 9.01 $2s^2$											5 B Boron 10.81 $2s^22p^1$	6 C Carbon 12.01 $2s^22p^2$	7 N Nitrogen 14.01 $2s^22p^3$	8 O Oxygen 16.00 $2s^22p^4$	9 F Fluorine 19.00	10 Ne Neon 20.18
3	11 Na Sodium 22.99 $3s^1$	12 Mg Magnesium 24.31 $3s^2$											13 Al Aluminum 26.98 $3s^23p^1$	14 Si Silicon 28.09 $3s^23p^2$	15 P Phosphorus 30.97 $3s^23p^3$	16 S Sulfur 32.07 $3s^23p^4$	17 Cl Chlorine 35.45	18 Ar Argon 83.90
4	19 K Potassium 39.10 $4s^1$	20 Ca Calcium 40.08 $4s^2$	21 Sc Scandium 44.96 $4s^23d^1$	22 Ti Titanium 47.87 $4s^23d^2$	23 V Vanadium 50.94 $4s^23d^3$	24 Cr Chromium 52.00 $4s^13d^5$	25 Mn Manganese 54.94	26 Fe Iron 55.85 $4s^23d^6$	27 Co Cobalt 58.93 $4s^23d^7$	28 Ni Nickel 58.69 $4s^23d^8$	29 Cu Copper 63.54 $4s^13d^{10}$	30 Zn Zinc 65.39 $4s^23d^{10}$	31 Ga Gallium 67.72 $4s^24p^1$	32 Ge Germanium 72.61 $4s^24p^2$	33 As Arsenic 74.92 $4s^24p^3$	34 Se Selenium 78.96 $4s^24p^4$	35 Br Bromine 79.90	36 Kr Krypton 83.80
5	37 Rb Rubidium 85.469 $5s^1$	38 Sr Strontium 87.82 $5s^2$	39 Y Yttrium 88.908 $5s^24d^1$	40 Zr Zirconium 91.22 $5s^24d^2$	41 Nb Nobelium 92.906 $5s^14d^4$	42 Mo Molybdenum 95.94 $5s^14d^4$	43 Tc Technetium (97)	44 Ru Ruthenium 102.91 $5s^14d^7$	45 Rh Rhodium 102.91 $5s^14d^8$	46 Pd Palladium 106.4 $4d^{10}$	47 Ag Silver 107.87 $5s^14d^{10}$	48 Cd Cadmium 112.4 $5s^24d^{10}$	49 In Indium 114.82 $5s^25p^1$	50 Sn Tin 116.69 $5s^25p^2$	51 Sb Antimony 121.75 $5s^25p^3$	52 Te Tellurium 127.6 $5s^25p^4$	53 I Iodine 126.9	54 Xe Xenon 131.3
6	55 Cs Cesium 132.91 $6s^1$	56 Ba Barium 137.33 $6s^2$	71 Lu Lutetium 174.97 $6s^25d^1$	72 Hf Hafnium 172.49 $6s^25d^2$	73 Ta Tantalum 190.95 $6s^25d^3$	74 W Tungsten 183.85 $6s^25d^4$	75 Re Rhenium 186.21	76 Os Osmium 190.2 $6s^25d^6$	77 Ir Iridium 192.22 $6s^25d^7$	78 Pt Platinum 195.09 $6s^15d^9$	79 Au Gold 196.97 $6s^15d^{10}$	80 Hg Mercury 200.59 $6s^25d^{10}$	81 Tl Thallium 204.37 $6s^26p^1$	82 Pb Lead 207.2 $6s^26p^2$	83 Bi Bismuth 208.96 $6s^26p^3$	84 Po Polonium (208) $6s^26p^4$	85 At Astatine (210)	86 Rn Radon (222)
7	87 Fr Francium (223) $7s^1$	88 Ra Radium 226.03 $7s^2$	103 Lr Lawrencium (260) $7s^26d^1$	104 Unq Unnilquadium (261) $7s^26d^2$	105 Unp Unnilpentium (262) $7s^26d^3$	106 Unh Unnilhexium (263) $7s^26d^4$	107 Uns Unnilseptium (262)	108 Uno Unniloctium (265) $7s^26d^6$	109 Uun Ununilium (269) $7s^26d^7$	110 Uun Ununilium (269) $7s^26d^7$	111 Uuu Unununium (272)	112 Uub						

f-block

8	57 La Lanthanum 138.91 $6s^25d^1$	58 Ce Cerium 140.12 $6s^25d^14f^1$	59 Pr Praseodymium 140.907 $6s^24f^3$	60 Nd Neodymium 144.24 $6s^24f^4$	61 Pm Promethium (145) $6s^24f^5$	62 Sm Samarium 150.36 $6s^24f^6$	63 Eu Europium 151.965 $6s^24f^7$	64 Gd Gadolinium 157.25 $6s^25d^14f^7$	65 Tb Terbium 158.93 $6s^24f^9$	66 Dy Dysprosium 162.50 $6s^24f^{10}$	67 Ho Holmium 164.93 $6s^24f^{11}$	68 Er Erbium 167.26 $6s^24f^{12}$	69 Tm Thulium 168.93 $6s^24f^{13}$	70 Yb Ytterbium 173.04
9	89 Ac Actinium (227) $7s^26d^1$	90 Th Thorium 232.04 $7s^26d^2$	91 Pa Protactinium 231.03 $7s^26d^15f^2$	92 U Uranium 238.029 $7s^26d^15f^3$	93 Np Neptunium (237) $7s^26d^15f^4$	94 Pu Plutonium (244) $7s^25f^6$	95 Am Americium (243) $7s^25f^7$	96 Cm Curium (247) $7s^26d^15f^7$	97 Bk Berkelium (247) $7s^25f^9$	98 Cf Californium (251) $7s^25f^{10}$	99 Es Einsteinium (254) $7s^25f^{11}$	100 Fm Fermium (257) $7s^25f^{12}$	101 Md Mendelevium (258) $7s^25f^{13}$	102 No Nobelium (259) $7s^25f^{14}$

Now here are some problems to review what you have briefly learned about the Periodic Table of the Elements:

(1) What is the chemical symbol for the element Hydrogen (clue: it is in column 1, row 1); what is Hydrogen's Atomic Number (the number of protons in its nucleus); what is Hydrogen's Average Atomic Mass; what is Hydrogen's physical state at Standard Temperature and Pressure (Gas, Liquid or Solid); Is Hydrogen a Metal, a Semi-Metal or a Non-metal; is Hydrogen's outermost electrons in a s-orbital, a p-orbital, a d-orbital, or an f-orbital; and is Hydrogen a member of the Alkali Metal Family?

(2) What is the chemical symbol for the element Potassium (clue: it is in column 1, row 4); what is Potassium's Atomic Number (the number of protons in its nucleus); what is Potassium's Average Atomic Mass; what is Potassium's physical state at Standard Temperature and Pressure (Gas, Liquid or Solid); Is Potassium a Metal, a Semi-Metal or a Non-metal; is Potassium's outermost electrons in a s-orbital, a p-orbital, a d-orbital, or an f-orbital; and is Potassium a member of the Alkali Metal Family?

(3) What is the chemical symbol for the element Calcium (clue: it is in column 2, row 4); what is Calcium's Atomic Number (the number of protons in its nucleus); what is Calcium's Average Atomic Mass; what is Calcium's physical state at Standard Temperature and Pressure (Gas, Liquid or Solid); Is Calcium a Metal, a Semi-Metal or a Non-metal; are Calcium's outermost electrons in a s-orbital, a p-orbital, a d-orbital, or an f-orbital; and is Calcium a member of the Alkali Earth Metal Family?

(4) What is the chemical symbol for the element Titanium (clue: it is in column 4, row 4); what is Titanium's Atomic Number (the number of protons in its nucleus); what is Titanium's Average Atomic Mass; what is Titanium's physical state at Standard Temperature and Pressure (Gas, Liquid or Solid); Is Titanium a Metal, a Semi-Metal or a Non-metal; are Titanium's outermost electrons in a s-orbital, a p-orbital, a d-orbital, or an f-orbital; and is Titanium a member of the Transition Elements?

(5) What is the chemical symbol for the element Cesium (clue: it is in column 5, row 8); what is Cesium's Atomic Number (the number of protons in its nucleus); what is Cesium's Average Atomic Mass; what is Cesium's physical state at Standard Temperature and Pressure (Gas, Liquid or Solid); Is Cesium a Metal, a Semi-Metal or a Non-metal; are Cesium's outermost electrons in a s-orbital, a p-orbital, a d-orbital, or an f-orbital; and is Cesium a member of the lanthanide elements?

(6) What is the chemical symbol for the element Uranium (clue: it is in column 7, row 9); what is Uranium's Atomic Number (the number of protons in its nucleus); what is Uranium's Average Atomic Mass; what is Uranium's physical state at Standard Temperature and Pressure (Gas, Liquid or Solid); Is Uranium a Metal, a Semi-Metal or a Non-metal; are Uranium's outermost electrons in a s-orbital, a p-orbital, a d-orbital, or an f-orbital; and is Uranium a member of the actinide elements?

(7) What is the chemical symbol for the element Aluminum (clue: it is in column 13, row 3); what is Aluminum's Atomic Number (the number of protons in its nucleus); what is Aluminum's Average Atomic Mass; what is Aluminum's physical state at Standard Temperature and Pressure (Gas, Liquid or Solid); Is Aluminum a Metal, a Semi-Metal or a Non-metal; are Aluminum's outermost electrons in a s-orbital, a p-orbital, a d-orbital, or an f-orbital; and what chemical family is Aluminum a member of (Main group elements, Halogens, Rare Gases)?

(8) What is the chemical symbol for the element Carbon (clue: it is in column 14, row 2); what is Carbon's Atomic Number (the number of protons in its nucleus); what is Carbon's Average Atomic Mass; what is Carbon's physical state at Standard Temperature and Pressure (Gas, Liquid or Solid); Is Carbon a Metal, a Semi-Metal or a Non-metal; are Carbon's outermost electrons in a s-orbital, a p-orbital, a d-orbital, or an f-orbital; and what chemical family is Carbon a member of (Main group elements, Halogens, Rare Gases)?

(9) What is the chemical symbol for the element Nitrogen (clue: it is in column 15, row 2); what is Nitrogen's Atomic Number (the number of protons in its nucleus); what is Nitrogen's Average Atomic Mass; what is Nitrogen's physical state at Standard Temperature and Pressure (Gas, Liquid or Solid); Is Nitrogen a Metal, a Semi-Metal or a Non-metal; are Nitrogen's outermost electrons in a s-orbital, a p-orbital, a d-orbital, or an f-orbital; and what chemical family is Nitrogen a member of (Main group elements, Halogens, Rare Gases)?

(10) What is the chemical symbol for the element Oxygen (clue: it is in column 16, row 2); what is Oxygen's Atomic Number (the number of protons in its nucleus); what is Oxygen's Average Atomic Mass; what is Oxygen's physical state at Standard Temperature and Pressure (Gas, Liquid or Solid); Is Oxygen a Metal, a Semi-Metal or a Non-metal; are Oxygen's outermost electrons in a s-orbital, a p-orbital, a d-orbital, or an f-orbital; and what chemical family is Oxygen a member of (Main group elements, Halogens, Rare Gases)?

(11) What is the chemical symbol for the element Chlorine (clue: it is in column 17, row 3); what is Chlorine's Atomic Number (the number of protons in its nucleus); what is Chlorine's Average Atomic Mass; what is Chlorine's physical state at Standard Temperature and Pressure (Gas, Liquid or Solid); Is Chlorine a Metal, a Semi-Metal or a Non-metal; are Chlorine's outermost electrons in a s-orbital, a p-orbital, a d-orbital, or an f-orbital; and what chemical family is Chlorine a member of (Main group elements, Halogens, Rare Gases)?

(12) What is the chemical symbol for the element Helium (clue: it is in column 18, row 1); what is Helium's Atomic Number (the number of protons in its nucleus); what is Helium's Average Atomic Mass; what is Helium's physical state at Standard Temperature and Pressure (Gas, Liquid or Solid); Is Helium a Metal, a Semi-Metal or a Non-metal; are Helium's outermost electrons in a s-orbital, a p-orbital, a d-orbital, or an f-orbital; and what chemical family is Helium a member of (Main group elements, Halogens, Rare Gases)?

(13) What is the chemical symbol for the element Neon (clue: it is in column 18, row 2); what is Neon's Atomic Number (the number of protons in its nucleus); what is Neon's Average Atomic Mass; what is Neon's physical state at Standard Temperature and Pressure (Gas, Liquid or Solid); Is Neon a Metal, a Semi-Metal or a Non-metal; are Neon's outermost electrons in a s-orbital, a p-orbital, a d-orbital, or an f-orbital; and what chemical family is Neon a member of (Main group elements, Halogens, Rare Gases)?

Congratulations! To answer these questions you had to make extensive use of the Periodic Table of the Elements. As you continue to use this textbook you will come to use it with an even greater degree of skill.

Chapter 2
Trying to Understand Matter: From the Ancient World to the 1800's

   Now let us review the history of Chemistry and how humankind came to learn so much about the elements in the Periodic Table. Basically our understanding about the nature of any type of matter has been limited by our ability to break down chemicals into pure elements. Later, people were limited by their abilities to break down the individual atoms of the elements into their sub-atomic components (electrons, protons, and neutrons) and, subsequently, to break down the electrons, protons, and neutrons even further.

   Around 1000 BC, a Hebrew writer described much about the ancient search for and knowledge of different forms of matter. These were metaphors for Job's struggle to gain the wisdom and understanding needed to understand all of the adversity he was experiencing (28th Chapter of Job (New International Version)):

   "There is a mine for silver and a place where gold is refined. Iron is taken from the earth, and copper is smelted from ore. Man puts an end to the darkness; he searches the farthest recesses for ore in the blackest darkness. Far from where people dwell he cuts a shaft, in places forgotten by the foot of man; far from men he dangles and sways. The earth, from which food comes, is transformed below as by fire; sapphires come from its rocks, and its dust contains nuggets of gold. No bird of prey knows that hidden path, no falcon's eye has seen it. Proud beasts do not set foot on it, and no lion prowls there. Man's hand assaults the flinty rock and lays bare the roots of the mountains. He tunnels through the rock; his eyes see all its treasures. He searches the sources of the rivers and brings hidden things to light. 'But where can wisdom be found?' 'Where does understanding dwell?' Man does not comprehend its worth; it cannot be found in the land of the living. The deep says, 'It is not in me'; the sea says, 'It is not with me.' It cannot be bought with the finest gold, nor can its price be weighed in silver. It cannot be bought with the gold of Ophir, with precious onyx or sapphires. Neither gold nor crystal can compare with it, nor can it be had for jewels of gold. Coral and jasper are not worthy of mention; the price of wisdom is beyond rubies. The topaz of Cush cannot compare with it; it cannot be bought with pure gold. "Where then does wisdom come from? Where does understanding dwell? It is hidden from the eyes of every living thing, concealed even from the birds of the air. Destruction and Death say, 'Only a rumor of it has reached our ears.' God understands the way to it and he alone knows where it dwells, for he views the ends of the earth and sees everything under the heavens. When he established the force of the wind and measured out the waters, when he made a decree for the rain and a path for the thunderstorm, then he looked at wisdom and appraised it; he confirmed it and tested it. And he said to man, 'The fear of the Lord—that is wisdom, and to shun evil is understanding.' "

   So let us seek knowledge of Chemistry together, hoping to grow in wisdom and understanding too. In the fifth century BC (about 400 years before the current era), Greek philosopher Empedocles hypothesized that all matter was composed of four basic elements: earth, air, fire and water. At about the same time, the Greek philosopher Deocritus had a theory that all matter was composed of unchangeable atoms of the elements which he thought were in continuous random motion. The now well known Greek philosopher Plato later on adopted Empedocles' theory of the composition of matter and his pupil, Aristotle, later, a well respected philosopher in his own right, adopted Plato's theory about matter's make-up. The Romans accepted their teachings too. With the collapse of the Roman Empire, Christians in the Dark Ages and Middle Ages also accepted them.

As can be seen from the quote from the book of Job, way before this theorizing of the makeup of matter, humans began to create the metal alloys of bronze and the pure elemental metal iron in what historians call the Bronze and Iron Ages. They heated up mineral ores to extract, without fully realizing the significance of what they were doing, the elements of copper and tin (to make bronze) and iron. Much mining in ancient times extracted precious metals like gold and silver and then purifying them through a process called refining. Semi-precious and precious gemstones were also extracted from the ground and cut using a crude knowledge of these gemstone crystal's structures. Of course food preparation, especially baking, was going on at the same time with its own unique food chemistry. And let us not forget color chemistry that entered into ancient art, fabric dyes, paint, and cosmetics. And naturally medical chemistry's use of herbs, plant materials, and minerals is quite ancient. So is the brewing of beers and the making of wines. As people changed mater and extracted different substances from the earth they really got to thinking about what rules governed changing matter and what made up matter.

Aristotle believed that the four elements, of which Emedocles hypothesized all matter to be composed, each had two qualities by which they could be converted into another element. Air and fire both had the quality of heat that allowed conversion between them. Air and water had the quality of wetness by which they could be converted to the other. Water and earth had the quality of coldness by which they could be converted to the other. And earth and fire shared the property of dryness that allowed them to be converted back and forth. But Aristotle and the other Greek philosophers left their hypotheses at just thinking. You will find many centuries later that Chemistry has become an experimental science that tests its hypotheses. But the ancients were not doing much testing of their hypotheses.

As recently as the 1600's (the seventeenth century), the metals prevalent in Western countries were thought to simply be forms of Aristotle's "earth" element (with the exception of liquid mercury). By that time the ideas of "alchemy " had developed. Its central concept was that metals "mature" in the earth, beginning with dull and dirty lead and maturing in the end as bright and shiny gold. Since all metals shared the qualities of shininess, denseness, and malleability in common, they were thought of by alchemists as differing only in degree, but were not completely different from one another. In that century English playwright William Shakespeare's *King Lear* goes refers frequently to the concepts of alchemy. Alchemists tied in vain to convert lead into gold. Eventually alchemists started working with the different forms of matter. Alchemists burnt, distilled, melted and condensed an amazing array of different substances.

In the seventeenth century Briton Sir Robert Boyle who was a friend of Sir Isaac Newton wrote the book *The Sceptical Chymist* (1661). In it he gave alchemists a nod and a wink while disparaging many of their attempts to convert lead into gold. He argued that the elements would not be found by simply theorizing but by experimentation. Boyle believed that elements must be "certain primitive and simple, or perfectly unmingled bodies which not being made of any other bodies, or of one another, are the ingredients of which all those called perfectly mixt bodies are immediately compounded, and into which they are ultimately resolved."[1]This is close to the modern concept of an element.

And a hundred years later, Frenchman Antoine Laurent Lavoisier in 1783 concluded from his experiments with water: "it is not a simple substance at all (as Aristotle and the alchemists thought), not properly called an element."[2] He used electricity to separate the elements that make up water (hydrogen and oxygen). He named the constituents of water hydrogen (water-former) and oxygen (acid-former). He found that they combine in a two-to-one ratio reflected by the chemical formula H_2O.

Sometimes chemists' use of the term "element" is confusing because they use it interchangeably to describe either a specific kind of atom (remember the Periodic Table of the Elements) or a physical substance containing only one kind of atom (like oxygen gas, O_2, or nitrogen gas, N_2).

In the eighteenth century (the 1700s) came "pneumatick chemistry." It concentrated on the physical properties of gases (airs). English clergyman Stephen Hales developed in the early part of the 1700's a pneumatic trough for collecting the gases that came out of heated substances. Englishman Jospeh Priestly made numerous discoveries with Hales' pneumatic trough. Priestly isolated "about twenty different gases (airs) including hydrogen chloride, nitric oxide, and ammonia."[6] But Priestly, and other "pneumatick" chemists like Scotsman Joseph Black and his student, Daniel Rutherford, regarded the different substances that they isolated with Hales' trough as "common air" that had been modified in some manner. So they were sticking to the Aristotelian concepts of four elements: air, water, earth, and fire.

Next came one last wrong turn when chemists thought up the idea of "phlogiston." Phlogiston has been attributed to Arab Jabir ibn Hayyan's *real sulfur* that is a combustible "yellow solid mined from the Earth, a component of gunpowder, and the brimstone that bubbles beneath the fires of hell."[7] George Ernst Stahl (1660-1734) gave it the name "phlogiston" from the Greek word "to burn". "To some chemists phlogiston was fire itself: a form of the ancient element (fire)." "It seems reasonable enough to assume from the flames and smoke dancing above a burning log that the wood is releasing some substance into the air. This, then, was phlogiston, the essence of flammability."[8]

In the 18th century (the 1700's) when metals were heated in the presence of air, called calcination, they were converted to dull "calxes" and could be converted back to the original metals by heating the "calxes" in the presence of charcoal. Presbyterian minister, Joseph Priestly and his patron, the Earl of Shelborne, heated the "calx of mercury" (mercuric oxide) which converted it to mercury with the release of oxygen. Antoine Lavosier learned of this experimental result from French pharmacist Pierre Bayen. Bayen also informed Lavoisier that no charcoal had to be present.

Priestly also had collected the gaseous product of heating the mercuric oxide in Hales' pneumatic trough. He found that mice survived longer breathing this gas, this "pure air," this "dephlogisticated air" (which we now know to be oxygen) than when they breathed just normal air. After having dinner with Priestly and the Earl of Shelborne, all of this lead Lavoisier in March 1775 to announce that "his own experiments with mercuric oxide revealed all calxes to be a combination of metals with such a gas."[9] "Lavoisier came to understand that this "pure air" was a substance in its own right (the element oxygen)."[10]

And concerning the Aristotelian element, air, Lavoisier concluded from his experiments that "it was comprised of two elastic fluids of different and opposite qualities."[3]
Lavoisier found that the "highly respirable" elastic fluid part of air was an element in and of itself, oxygen (acid former).[4] The other elastic fluid in air was inimical to life he called by the Greek word, azot, but he also that it formed part of nitric acid and could therefore be called "nitrigen"[5] Englishman Henry Cavendish also isolated the main components of air.

Lavoisier also discovered that "many non-metallic elements such as sulfur, carbon and phosphorous, combine with oxygen to produce gases that dissolve in water to make acids, and that is why Lavoisier named the new element as he did (in German oxygen is still known as *Sauerstuff*, "acid stuff")."[11] This was a misunderstanding about acids which we will clear up later.

"The discovery of oxygen did not make phlogiston redundant; the two are fundamentally incompatible."[12] So in 1785 Lavoisier published his formal denunciation of phlogiston: "Chemists have made phlogiston a vague principle, which is not strictly defined and which consequently fits all the explanations demanded of it."[13] In 1789 Lavoisier published a textbook, *An Elementary Treatise on Chemistry*, that consolidated his oxygen theory and "defined an element as any substance that could not be split into simpler components by chemical reactions."[14] This is the same definition for an element that we use in the 21st century.

At the turn of the 19th century (1800) Manchester England Quaker John Dalton began drawing pictures of atoms, giving the notion of atoms " a concrete expression, and … this helped make sense of chemists ' analyses of the composition of matter."[15] "Dalton had every confidence in the 'solid, massy, hard impenetrable, movable particles' that (were the atoms that) Isaac Newton envisaged over a hundred years earlier. Dalton imagined them as eternal, unchangeable bodies, however inaccessible to the human eye."[16] Dalton was visualizing Greek philosopher Democritus's atoms. While Dalton did not know what atoms were made of, he theorized that the weights of atoms were identical for atoms of the same element but differ for different elements. So Dalton's atomic theory "allowed chemistry to become an exact science." "If atoms were little balls that always united in the same simple ratios to make 'compound particles,' this explained why chemical reactions between elements always took place in constant and simple proportions."[17] "The importance of making numerically precise measurements of chemical processes had been clear enough to Cavendish, Priestly, and their contemporaries, but without an underlying theory of the elements, these numbers were merely codification's of empirical observations."[18]

Dalton published his atomic theory in a book, *A New System of Chemical Philosophy*, in 1808. But Dalton used hieroglyphic symbols for the elements in his book. Several years later Berzelius proposed that "they be replaced with an alphabetic notation for the elements. Berzelius had the idea that one could represent each element by the first letter of its name; or in cases where this did not uniquely distinguish them from others, by two letters. Thus, hydrogen becomes H, oxygen O, and carbon C; Cobalt was to be distinguished from carbon by the designation Co." "Berzelius was insisted that Latin names should be used for the elements that possessed them; so copper becomes Cu (*cuprum*), gold is Au (*aurum*), and iron is Fe (*ferrum*)."[19]

Dalton's pictures of atoms and molecules provided a unification of the visible (macro) and invisible (micro) worlds of Chemistry. They "show at once what we can observe (for example, hydrogen and oxygen combining to make water) and what we cannot: the union of real, tangible atoms."[20]

Russian Dimitri Mendeleyev was highly successful in the 1800's at organizing the known elements, more successful than even Dalton. Mendeleyev vindicated the published works of English chemist John Newlands. Newlands even gave a talk in 1866 to the Royal Society stating that "if the elements were published in order of atomic mass, (then) each element shared properties with those eight and sixteen places later. In other words, the properties repeated periodically every eight elements. Newlands drew an analogy to music, wherein each scale begins afresh every eith notes (an octave pattern)."[21]

"Mendeleyev's insight was not so much to find order amongst the elements but to find the order that underlay the (known 60) elements. The difference between these two concepts is evident in the way that Mendeleyev's Periodic Table of the Elements (published in 1869) leaves gaps, some with a question mark inserted."[21] "He used the periodic trends that the Table embodied, to predict, in some detail, the properties of the missing elements."[22]

So we have followed mankind's understanding of matter from ancient times all of the way to Mendeleyev's Periodic Table. It later evolved into that which was presented in Chapter 1 as the modern Periodic Table of the Elements.

Chapter 3
Chemistry Determines the Basic Components and Terminology of the Atom

    While John Dalton was not so concerned with the internal structure of the atom, others were: Englishman J.J. Thomson and Ernest Rutherford. In the 1890's J.J. Thomson performed experiments with a long glass tube with a vacuum inside. He applied a voltage across two wires (electrodes) at either end of the long vacuum tube. He observed a glowing beam traveling from the negatively charged electrode (the cathode) to the positively charged electrode (the anode). He called the beam cathode rays since it began at the cathode and went to the anode. Thomson did further experiments on the cathode rays and was able to change their path with a magnet placed at the center of the vacuum tube while a voltage was applied across the electrodes. From this observation he concluded that the cathode rays contained particles with negative charges. He called these particles electrons. Even further experiments led Thomson to conclude that electrons were very small and light. He used these experimental findings to create a "plum-pudding" model of the internal structure of atoms: atoms' internal structure was of a positively charged pudding with small negatively charged electrons scattered about the positively charged pudding.

    In 1904 Ernest Rutherford decided that one way to learn about the internal structure of atoms was to smash high-energy particles into them. So he used a very thin sheet of gold which can be hammered tissue thin (only several atoms thickness). "'I was brought up to look at the atom as a nice hardy fellow, red or grey in color, according to taste, ' Rutherford once said. But in 1907 he found that atoms were not so hard at all. They were mostly empty space. Working at Manchester University in England, Rutherford and his students Hans Geiger and Ernest Marsden fired positively charged alpha particles from radioactive elements at thin gold foil, and found that the particles could 'see' right through the ponderous (gold) element. Mostly they passed through the foil with scarcely a deflection in their course."[23]

    But an even more surprising result was when some of the positively charged alpha particles bounced back at Rutherford off the gold foil. Rutherford described his amazement at this discovery as being the same as if he had fired a cannon at a tissue and the cannon ball bounced back at him off the tissue. Rutherford, being a good scientist thought about the implications of this observation. He concluded that it meant that atoms where largely composed of empty space with small solid parts. Rutherford, like Thomson, put forth a model for the internal structure of element's atoms: the "nuclear" model of atoms: each atom had a small positively charged nucleus at its center surrounded by a cloud of rapidly orbiting, negatively charged electrons. When the positively charged alpha particles made a direct hit on the positively charged nuclei of the gold atoms in Rutherford's foil, they were repelled (like charges repel). When they missed the positively charged nuclei of the gold foil's atoms, they passed through, sometimes having their paths be slightly charged by the positive charges of the nuclei.

    Further experimentation led to the conclusion that atoms were composed of three types of particles: positively charged protons in the nucleus, neutral particles in the nucleus called neutrons, and negatively charged electrons orbiting around the nucleus. These additional experiments led scientists to conclude that the mass of the protons and neutrons were almost the same. A proton's weight and a neutron's weight was set at one atomic mass unit (1 amu) apiece. 1 amu was found experimentally to be = 9.11×10^{-27} kg. And the electron's masses were found to be to equal 1/1836 amu or essentially 0 amu.

The number of protons in the nucleus of an atom of an element is always the same. The number of protons in the nucleus of an atom always determines what element it is. This starts modern atomic terminology (nomenclature): the number of protons in the nucleus of an atom of an element is its atomic number. Remember the whole numbers on top of each atomic symbol in the periodic table in Chapter 1? They are the Atomic Numbers. The number of neutrons in the nuclei of atoms of an element can vary. More atomic Nomenclature is that the different numbers of neutrons in the nuclei of atoms of an element create different isotopes of the element. Isotopes of an element have the same number of protons in their atoms' nuclei, but have different numbers of neutrons.

The sum of an atom's number of protons and its number of neutrons is that atom's atomic mass. As mentioned in the previous paragraph, elements can different number of neutrons in their atoms' nuclei. These atoms of the same element (having the same number of protons) with different numbers of neutrons in their nuclei are the different isotopes of an element. To arrive at the average atomic mass of an element, a weighted average of the atomic masses of its different isotopes is determined. To give a formula for this weighted average for an element, we need to introduce the symbols (they're like mathematical variables) for atomic number, Z (the number of protons), atomic mass, A (the number of neutrons plus the number of protons), for any element X whose chemical symbol appears in the periodic table in Chapter 1:

$$^{A}_{Z}X$$

The Atomic Mass for ^{1}H is one amu. The atomic mass for ^{19}F is 19 amu. Boron has two isotopes, ^{10}B (5 protons and 5 neutrons in its nucleus) and ^{11}B (5 protons and 6 neutrons). The atomic mass of the isotope ^{10}B is 10.013 amu and its percentage abundance is 19.9% (.199) and the mass of the isotope ^{11}B is 11.009 amu and its percentage abundance is 80.1% (.801). So the average atomic mass of Boron is computed as the weighted average of 10.013 x 0.199 + 11.009 x .801 = 10.81 amu.

Chapter 4
The Internal Structure of the Atoms of Elements

Dutch Scientist Niels Bohr was able to study the light emitted from a glass tube filled with the element Hydrogen as he exposed the Hydrogen to a wide range of energy. The Hydrogen did something surprising. A solid glass prism does when sunlight passes through it it radiates out a continuous spectrum of light of all visible frequencies and wavelengths (a rainbow). Hydrogen in a sealed glass tube, instead, gives off light (it emits light) only of very specific wavelengths and frequencies in what is called a line spectrum:

Bohr wondered what could be causing these specific lines of Hydrogen's emission spectrum instead of a continuous emission spectrum as a glass prism does to sunlight. He came up with the "planetary" or "solar system" model of the atom to explain Hydrogen's line emission spectrum. In the Bohr's planetary model of the atom, electrons travel around each atoms' nucleus in circular orbits similar to the elliptical orbits in which the planets in our solar system travel around the Sun. Just the gravitational attractions between the Sun and the planets hold the planets in their orbits, so the electrical attractions between the negatively charged electrons and the positively charged nucleus in the Bohr model hold the electrons near the nucleus in paths called *orbitals*.

Bohr's model had the provision that the farther away from the nucleus that an electron was, the more energy it had. The energies of the electrons specifically were determined by the order of the orbitals surrounding atoms' nuclei. Bohr used the symbol or variable n to hold the values for the order of the electron orbitals. The first electron orbital, being the one closest to the nucleus, was assigned a value of $n=1$. The second electron orbital, being the second closest electron orbital to the nucleus, was assigned a value of $n=2$ and so forth.

Bohr believed that when an electron was in the innermost $n=1$ orbital of its atom, it was in its ground state. And when that electron absorbed energy, it enters into an excited state and goes to one of the higher order orbitals ($n=2,3,4,...$). Eventually, an electron in its excited state loses energy, given off as light of a specific frequency or wavelength, and drops back to its ground state in the $n=1$ orbital.

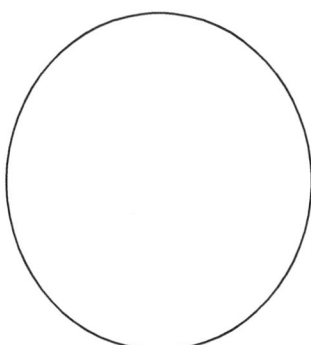

The light given off by an electron as it goes from one of its excited states, in an outer orbital with n=2,3,4,5…, to its ground state has an energy equal to the difference in energy between the excited state and the ground state. This difference in energy is represented by the equation E=$nh\nu$ where n=2,3,4,5,… (the excited state electron orbital number), h=Plank's constant, and ν is the frequency of the light emitted. The relationship between light wavelength, λ, and light frequency, ν, is given by the equation λ=1/ν.

Notice that in Bohr's equation the energy given off when an electron goes from one of its excited states to its ground state is directly related to the frequency of the light emitted. Bohr found his equation correctly predicted the frequencies of the line spectrum of the hydrogen atom that he observed. He was awarded the 1922 Nobel Prize in Physics for his discovery.

Chapter 5
Quantum Mechanics and Chemistry

Niels Bohr's equation correctly predicted the element Hydrogen's orbital energies and the energies and frequencies of light emitted when different energies are applied to the tube of hydrogen gas. But it didn't correctly predict the energies of other element's orbitals or the frequency of light emitted as their electrons in excited state in the higher orbitals went to their ground state. A new model of elements' atoms called the quantum mechanics model was developed to explain our current understanding of how atoms' internal structures.

Chapter 6

Scratchpad of Stoichiometry Problems

Stoichiometry Problems

(A) Calculating Molar Masses (These will be used in Stoichiometry Calculations)

Always use the Element's(s') Atomic Mass(es) from the Periodic Table to calculate molar masses. If an element(s) chemical formula has any subscripts, for example, O_2 or H_2 or H_2SO_4; then multiply the atomic mass(es) of the element(s) by those subscripts. Do not multiply the atomic mass(es) of element(s) by coefficients, for example, $3H_2 + N_2 \rightarrow 2NH_3$.

(1) Calculating the Molar Mass of a Monoatomic Element.

(a) Example: Find the Molar Mass of the monoatomic element Copper (Cu) in the Balanced Chemical Equation $Cu + 2AgNO_3 \rightarrow 2Ag + Cu(NO_3)_2$

Answer: The atomic mass of Cu (Column 1B in the Periodic Table) is 63.55. So the molar mass of the monoatomic element Copper in the reaction: $Cu + 2AgNO_3 \rightarrow 2Ag + Cu(NO_3)_2$ is $\underline{63.55 \text{ grams Cu}}$.
 1 mole Cu

(b) Practice Problem: What is the molar mass of monoatomic element Silver (Ag) as in the reaction: $Cu + 2AgNO_3 \rightarrow 2Ag + Cu(NO_3)_2$?

Answer: $\underline{107.8 \text{ g Ag}}$
 1 mol Ag

(2) Calculating the Molar Mass of a Diatomic Element.

(a) Example: Find the molar mass of the diatomic element O_2 (oxygen gas) as is found in the balanced chemical equation:
$C(s) + O_2 \rightarrow CO_2 (g)$:

Answer: The molar mass of O_2 = 2 (multiply by the subscript) x atomic mass of O (oxygen) = 2 x 15.99 = $\underline{31.98 \text{ grams } O_2}$
 1 mole O_2

(b) Practice Problem: What is the molar mass of diatomic element N_2?

Answer: $\underline{28 \text{ g } N_2}$
 1 mol N_2

(A)(3) Calculating the Molar Mass of a Polyatomic Chemical Compound
(either an Ionic Compound or a Molecular Compound)

(a) Example: What is the Molar Mass of Polyatomic Chemical Compound $AgNO_3$ in the Balanced Chemical Equation:
$Cu + 2AgNO_3 \rightarrow 2Ag + Cu(NO_3)_2$?

Answer: (Hint: Ignore the Equation's (the Reaction's) Coefficients. Multiply any Elements' Atomic Masses of $AgNO_3$ by their Subscripts. Then add the results together to get the Molar Mass of $AgNO_3$)

Molar Mass of $AgNO_3$ = Molar Mass of $Ag_1N_1O_3$ =

(Use the Multiplicative Identity Element of One with the Elements whose Chemical Formulas have no subscripts as normally written)

(1 x Atomic Mass Ag) + (1 x Atomic Mass N) + (3 x Atomic Mass of O)
= (1 x 107.9) + (1 x 14.0) + (3 x 16.0) = 107.9 + 14.0 + 96.0
= <u>169 grams $AgNO_3$</u>
   1 mole $AgNO_3$

Practice Problem: What is the Molar Mass of $Cu(NO_3)_2$?

Molar Mass of $Cu_1(NO_3)_2$ =

1 x Atomic Mass Cu + 2 x Atomic Mass N + (2 x 3 x Atomic Mass of O)
= 1 x 63.55 + 2 x 14.0 + 6 x 16.0
= 63.55 + 28.0 + 96.0
= <u>187 g $Cu(NO_3)_2$</u>
   1 mol $Cu(NO_3)_2$

(B) Determining Molar Ratios (These will be used in Stoichiometry Calculations)

(1) Example: List all the possible molar ratios in the balanced chemical equation:

$Cu + 2\ AgNO_3 = 2\ Ag + Cu(NO_3)_2 = 1xCu + 2xAgNO_3 = 2xAg + 1xCu(NO_3)_2$

(Use coefficients to construct molar ratios; ignore subscripts)
(Remember the Multiplicative Identity Element of One for Chemical Compounds without a coefficient in the Chemical Reaction as normally written: $Cu + 2\ AgNO_3 = 2\ Ag + Cu(NO_3)_2$.)

There are 12 possible molar ratios for this chemical reaction
$1xCu + 2xAgNO_3 = 2xAg + 1xCu(NO_3)_2$:

Molar Ratios are All possible combinations of Reactants to Products:

$\dfrac{1\ mol\ Cu}{1\ mol\ Ag}$, $\dfrac{1\ mol\ Cu}{1\ mol\ Cu(NO_3)_2}$, $\dfrac{2\ mol\ AgNO_3}{2\ mol\ Ag}$, $\dfrac{2\ mol\ AgNO_3}{1\ mol\ Cu(NO_3)_2}$

And All possible combinations of Reactants to Reactants:

$\dfrac{1\ mol\ Cu}{2\ mol\ AgNO_3}$, $\dfrac{2\ mol\ AgNO_3}{1\ mol\ Cu}$

And All possible combinations of Products to Reactants:

$\dfrac{2\ mol\ Ag}{2\ mol\ AgNO_3}$, $\dfrac{1\ mol\ Cu(NO_3)_2}{1\ mol\ Cu}$, $\dfrac{2\ mol\ Ag}{2\ mol\ Cu}$, $\dfrac{1\ mol\ Cu(NO_3)_2}{2\ mol\ AgNO_3}$

And All Possible Combinations of Products to Products:

$\dfrac{2\ mol\ Ag}{1\ mol\ Cu(NO_3)_2}$, $\dfrac{1\ mol\ Cu(NO_3)_2}{2\ mol\ Ag}$

(2) Practice Problem: List all possible Molar Ratios for the Balanced Chemical Reaction:

FeS + 2HCl → H_2S + $FeCl_2$

Use the Multiplicative Identity Element of One for Chemical Compounds without Coefficients:

1 x FeS + 2 x HCl → 1 x H_2S + 1 x $FeCl_2$

Create all possible Molar Ratios: (reactants to products):

$\dfrac{1 \text{ mol FeS}}{1 \text{ mol } H_2S}$, $\dfrac{1 \text{ mol FeS}}{1 \text{ mol } FeCl_2}$, $\dfrac{2 \text{ mol HCl}}{1 \text{ mol } H_2S}$, $\dfrac{2 \text{ mol HCl}}{1 \text{ mol } FeCl_2}$

(reactants to reactants):

$\dfrac{1 \text{ mol FeS}}{2 \text{ mol HCl}}$, $\dfrac{2 \text{ mol HCl}}{1 \text{ mol FeS}}$

{products to reactants)

$\dfrac{1 \text{ mol } H_2S}{1 \text{ mol FeS}}$, $\dfrac{1 \text{ mol } H_2S}{2 \text{ mol HCl}}$, $\dfrac{1 \text{ mol } FeCl_2}{1 \text{ mol FeS}}$, $\dfrac{1 \text{ mol } FeCl_2}{2 \text{ mol HCl}}$

(products to products):

$\dfrac{1 \text{ mol } H_2S}{1 \text{ mol } FeCl_2}$, $\dfrac{1 \text{ mol } FeCl_2}{1 \text{ mol } H_2S}$

(C) One Step Stoichiometry Problems

(1) Find the number of moles of a monoatomic element, given a certain number of grams of it. (Divide the grams of the monoatomic element by its molar mass: Which is to say: Multiply the grams of the monoatomic element by the reciprocal of its molar mass) (cancel the identical units):

(a) Example: How many moles of the monoatomic element Copper (Cu) are in 56 grams of Cu in the Balanced Chemical Equation:
$Cu + 2AgNO_3 = 2Ag + Cu(NO_3)_2$?

56 g Cu x $\dfrac{1}{\dfrac{63.55 \text{ g Cu}}{1 \text{ mol Cu}}}$ = 56 g Cu x $\dfrac{1 \text{ mol Cu}}{63.55 \text{ g Cu}}$ = $\dfrac{56 \text{ mol Cu}}{63.55}$

= 0.89 mol Cu (two significant digits) (See problem (A)(1)(a) for molar mass of Cu) (Notice that the reciprocal of the molar mass has g of Cu in the denominator and mol of Cu in the numerator. This ends up cancelling the g Cu out)

(b) Practice Problem: How many moles of the monoatomic element Silver (Ag) are in 56 g of Ag in the Balanced Chemical Equation: $Cu + 2AgNO_3 = 2Ag + Cu(NO_3)_2$?
(See (A)(1)(b) for Molar Mass of Ag):

56 g Ag	1 mol Ag	= 0.52 mol Ag (2 significant digits)
1	107.9 g Ag	

This is a stoichiometry matrix. The vertical lines (|) are multiplication operations and the horizontal lines (_) are division operations.

(2) Given the mass of a polyatomic compound (ionic or molecular), find the equivalent number of moles of that element.

(a) Example: What is the equivalent number of moles of $AgNO_3$ in 200 grams of it in the balanced chemical reaction $Cu + 2AgNO_3 = 2Ag + Cu(NO_3)_2$?

200 g $AgNO_3$	1 mol $AgNO_3$	= 1.17 mol $AgNO_3$
	169.9 g $AgNO_3$	

(D) Two Step Stoichiometry Problems

(1) Find the number of moles of a product from a balanced chemical reaction (gone to completion), given a certain mass (in grams) of a certain reactant.

(a) Example: How many moles of monoatomic element Silver, Ag, are produced when 56 grams of monoatomic element Copper, Cu, react with an excess of $AgNO_3$ in the balanced chemical equation:
$Cu + 2AgNO_3 = 2Ag + Cu(NO_3)_2$?

(Hint: First, use the calculation of the number of moles of Copper given a mass of 56g of monoatomic Copper in (C)(1)(a) by the molar ratio of monoatomic Silver to Monoatomic Copper in (B)(1) to find the number of moles of monoatomic Silver formed when the monoatomic Copper reacts with an excess of polyatomic compound Silver Nitrate $AgNO_3$ in the balanced chemical equation: $Cu + 2AgNO_3 = 2Ag + Cu(NO_3)_2$)

$$\frac{56g\ Cu}{\ } \mid \frac{1mol\ Cu}{63.55g\ Cu} \mid \frac{2mol\ Ag}{1mol\ Cu} = 1.76\ mol\ Ag$$

Practice Problem: How many moles of monoatomic Silver, Ag, are produced by the reaction of 56 grams of polyatomic compound Silver Nitrate, $AgNO_3$, with excess monoatomic Copper, Cu, in the balanced chemical reaction: $Cu + 2AgNO_3 = 2Ag + Cu(NO_3)_2$?

$$\frac{56g\ AgNO_3}{\ } \mid \frac{1mol\ AgNO_3}{169.9g\ AgNO_3} \mid \frac{2mol\ Ag}{2mol\ AgNO_3} =$$

                Reciprocal of   Molar ratio
                Molar mass   of Ag to
                Of $AgNO_3$ $AgNO_3$
                (A)(3)(a)   (B)(1)

(D) Two Step Stoichiometry Problems

(2) Finding the number of moles of a reactant used up in a balanced chemical reaction given a certain number of grams of a product produced by the consumption of that number of moles of reactant.

(a) Example: How many moles of reactant Cu are used up in the presence of excess $AgNO_3$ to produce 56 grams of product Ag in the balanced chemical reaction: $Cu + 2AgNO_3 = 2Ag + Cu(NO_3)_2$?

56g Ag | 1mol Ag | 1mol Cu   = 0.26mol Cu
         | 107.9g Ag | 2mol Ag

(b) Practice Problem: How many moles of $AgNO_3$ reacting with excess Cu are used up to produce 56 grams of Ag in the balanced chemical equation: $Cu + 2AgNO_3 = 2Ag + Cu(NO_3)_2$?

(3) Finding the number of moles of one reactant which will react with a known amount of a second reactant in a given balanced chemical reaction.

(a) Example: How many moles of Cu will react with 200 grams of $AgNO_3$ in the balanced chemical reaction: $Cu + 2AgNO_3 = 2Ag + Cu(NO_3)_2$?

200g $AgNO_3$ | 1mol $AgNO_3$ | 1mol Cu   = 0.59mol Cu
              | 169.9g $AgNO_3$ | 2mol $AgNO_3$

(b) Practice Problem: How many moles of $AgNO_3$ will react with 200 grams of Cu in the balanced chemical equation: $Cu + 2AgNO_3 = 2Ag + Cu(NO_3)_2$?

(D) Two Step Stoichiometry Problems

(4) Finding the number of moles of one product formed given a known number of grams of a second product produced in a given balanced chemical reaction.

(a) Example: How many moles of $Cu(NO_3)_2$ are formed if 200 grams of Ag are formed according to the balanced chemical reaction: $Cu + 2AgNO_3 = 2Ag + Cu(NO_3)_2$?

<u>200g Ag | 1mol Ag | 1mol $Cu(NO_3)_2$</u>   = 0.93mol $Cu(NO_3)_2$
           | 107.9g Ag | 2mol Ag

(b) How many moles of Ag are produced if 200 grams of $Cu(NO_3)_2$ are produced with the Ag in the balanced chemical reaction: $Cu + 2AgNO_3 = 2Ag + Cu(NO_3)_2$?

(5) Finding the mass in grams of a product produced given a certain number of moles of one reactant consumed according to a balanced chemical reaction.

(a) Example: How many grams of Ag are produced when 56 moles of Cu are consumed according to the balanced chemical reaction: $Cu + 2AgNO_3 = 2Ag + Cu(NO_3)_2$?

<u>56g Cu | 2mol Ag | 107.9g Ag</u> = 12,084g Ag = 1.21×10^4 g Ag
        | 1 mol Cu | 1mol Ag

(b) Practice Problem: How many grams of $Cu(NO_3)_2$ are produced when 56 moles of Cu react with excess $AgNO_3$ according to the balanced chemical equation: $Cu + 2AgNO_3 = 2Ag + Cu(NO_3)_2$?

(D) Two Step Stoichiometry Problems

(6) Finding the mass in grams of one reactant consumed in a balanced chemical reaction assuming that a certain number of moles of one product produced.

(a) Example: How many grams of Cu react (are used up) in the balanced chemical equation Cu + 2AgNO$_3$ = 2Ag + Cu(NO$_3$)$_2$, assuming that 56 moles of Ag are produced?

<u>56mol Ag</u> | <u>1mol Cu</u> | <u>63.5g Cu</u> = 1778g Cu = 1.78 x 10^{3}g Cu
          | 2mol Ag | 1mol Cu

(b) Practice Problem: How many grams of AgNO$_3$ are consumed to produce 56 grams of Ag in the balanced chemical reaction: Cu + 2AgNO$_3$ = 2Ag + Cu(NO$_3$)$_2$?

(7) Finding the number of grams of one reactant that will react with a known number of moles of another reactant given their balanced chemical equation.

(a) Example: How many grams of Cu will react with 200 moles of AgNO$_3$ in a chemical reaction with the balanced equation of Cu + 2AgNO$_3$ = 2Ag + Cu(NO$_3$)$_2$?
(Hint: this is the mirror image of (D)(3)(a) and (D)(3)(b))
<u>200mol AgNO$_3$</u> | <u>1mol Cu</u> | <u>63.55g Cu</u> = 6355g Cu = 6.35x10^{3}g Cu
                | 2mol AgNO$_3$| 1mol Cu

(b) Practice Problem: How many grams of AgNO$_3$ will react with 200 moles of Cu in the balanced chemical reaction: Cu + 2AgNO$_3$ = 2Ag + Cu(NO$_3$)$_2$?

(D) Two Step Stoichiometry Problems

(8) Finding the number of grams of one product produced in a known balanced chemical reaction in which a known number of moles of a second product are produced.

(a) Example: How many grams of $Cu(NO_3)_2$ are produced if 200 moles of Ag are also produced by a chemical reaction with the balanced equation of $Cu + 2AgNO_3 = 2Ag + Cu(NO_3)_2$?

Molar mass of $Cu(NO_3)_2$ =

1 x Cu's amu = 1 x 63.6 = 63.6

2 x N's amu = 2 x 14 = 28

+ 6 x O's amu = 6 x 16 = 96

$$\frac{187.6g\ Cu(NO_3)_2}{1mol\ Cu(NO_3)_2}$$

$$\frac{200mol\ Ag}{} | \frac{1mol\ Cu(NO_3)_2}{2mol\ Ag} | \frac{187.6g\ Cu(NO_3)_2}{1mol\ Cu(NO_3)_2} = 18,760g\ Cu(NO_3)_2 = 1.87 \times 10^4 g$$

(b) Practice Problem: How many grams of Ag are produced by the chemical reaction with the balanced equation $Cu + 2AgNO_3 = 2Ag + Cu(NO_3)_2$ if 200 moles of $Cu(NO_3)_2$ are also produced?

(E) Three Step Stoichiometry Problems

(1) Finding the number of grams of a certain product produced by the reaction of the certain mass of a reactant in a chemical reaction with a known chemical reaction with a balanced equation.

(a) Example: How many grams of Ag are produced when 56 grams of Cu react with excess $AgNO_3$ in a chemical reaction with balanced equation $Cu + 2AgNO_3 = 2Ag + Cu(NO_3)_2$?

$$\frac{56g\ Cu}{} | \frac{1mol\ Cu}{63.6g\ Cu} | \frac{2\ mol\ Ag}{1mol\ Cu} | \frac{107.9g\ Ag}{1mol\ Ag} = 190g\ Ag = 1.9 \times 10^2 g\ Ag$$

(b) Practice Problem: How many grams of Ag are produced by the reaction of 56 grams of $AgNO_3$ with excess Cu in a chemical reaction with the balanced equation $Cu + 2AgNO_3 = 2Ag + Cu(NO_3)_2$?

(E) Three Step Stoichiometry Problems

(2) Finding the number of grams of a reactant that are used up in a balanced chemical reaction if you know the mass of a product produced.

(a) Example: How many grams of Cu react to produce 56 grams of Ag in a chemical reaction with the balanced equation: $Cu + 2AgNO_3 = 2Ag + Cu(NO_3)_2$?

56g Ag | 1mol Ag | 1mol Cu | 63.6g Cu = 16.2g Cu = 1.62×10^1g Cu
       | 107.9g Ag | 2mol Ag | 1mol Cu

(b) Practice Problem: How many grams of $AgNO_3$ react to produce 56 grams of Ag in a chemical reaction with the balanced equation: $Cu + 2AgNO_3 = 2Ag + Cu(NO_3)_2$?

(3) Finding the number of grams of one reactant which will react with a known mass of a second reactant in a chemical reaction with a balanced equation.

(a) Example: How many grams of Cu will react with 200 grams of $AgNO_3$ in a chemical reaction with the balanced equation: $Cu + 2AgNO_3 = 2Ag + Cu(NO_3)_2$?

200g $AgNO_3$ | 1mol $AgNO_3$ | 1mol Cu | 63.6g Cu = 37,4g Cu
              | 169.9g $AgNO_3$ | 2mol $AgNO_3$ | 1mol Cu

(b) Practice Problem: How many grams of $AgNO_3$ will react with 200 grams of Cu in the chemical reaction with the balanced equation: $Cu + 2AgNO_3 = 2Ag + Cu(NO_3)_2$?

Chapter 7

Chemistry Labs

A. Combination Reactions of Metal Oxides with Water that Produce Metal Hydroxides:

1. Reactants and Products: Sodium Oxide (a metal oxide), Water (a molecule), and Sodium Hydroxide (a metal hydride –a strong base) - Na_2O, H_2O, NaOH. Chemical reaction equation: $Na_2O + H_2O \rightarrow$ NaOH. Balanced Equation: $Na_2O + H_2O \rightarrow 2NaOH$.

Na2O + H2O ⇌ 2NaOH

2 moles of Sodium oxide=   123.96 Grams

1 mole of Water=   18.02 Grams

	Na2O	H2O	2NaOH
Delta G			
Physical State	white, deliq	liquid	white, deliq

Materials:

Sodium Oxide (a metal oxide) - Na_2O

Distilled Water (a molecule) - H_2O

Two 150 milliliter beakers

25 milliliter graduated cylinder

Weighing glassine paper

Digital or triple beam balance

Celsius thermometer

pH Meter

Glass rod

Procedure:

Tare a piece of glassine weighing paper on a balance. Using the balance, weigh out 6 grams of sodium oxide on the paper. Label one of the (clean, dry, and empty) 150ml beakers with the characters Na_2O and weigh it on the balance. Record this as the empty Na_2O beaker mass. Transfer the 6 grams of sodium oxide into it. Weigh it again and record this as the filled Na_2O beaker mass. Pour 20 milliliters of distilled water into the 25ml graduate cylinder (be sure to measure from the bottom of the meniscus). Weigh the other (clean, dry) 150ml beaker when it is empty and record it as the empty water beaker mass. Decant (pour) the distilled water into it and weigh the 150ml beaker with the distilled water in it. Weigh it again on the balance and record it as the filled water beaker mass. Measure the temperature of the distilled water in its 150ml beaker with the Celsius thermometer. Record the water's temperature (call it the starting H_2O temperature). Measure the pH of the distilled water and record it (call it the starting H_2O pH). Be prepared to observe all changes to the sodium oxide when you add the 20ml of distilled water to it. Decant (pour) the 20ml of distilled water from its 150ml beaker to the 150ml beaker containing the 6gm of sodium oxide. Stir the resulting solution with a glass rod. Visually observe the changes that occur. Measure the resulting solution's temperature with the Celsius thermometer and its pH. Record your visual observations and measurements of temperature (final solution temperature) and pH (final solution pH) of the resulting solution. Weigh the resulting solution (the product and any unreacted reactants). Measure the volume of the resulting solution in the 25 ml graduated cylinder.

Discussion and Conclusion Questions:

Write the balanced equation of the reaction that you have carried out: $Na_2O + H_2O \rightarrow NaOH$. What type of reaction is this (combination, decomposition, substitution)? Does the temperature change indicate that an exothermic (heat was given off to the environment) or endothermic (heat was absorbed from the environment) reaction had occurred? Is the distilled water an acid, a base or neutral? Judging by the final pH of the resulting solution, is the product an acid or base? From the weight of the distilled water, what is the number of moles of it being used in this chemical reaction? From the weight of the sodium oxide, how many moles of it are being used in this chemical reaction? From the number of moles of the two reactants (the water and the sodium oxide), how many moles of sodium hydroxide theoretically could result (assuming a 100% yield)? From the mass of the final solution what % yield of NaOH appears to have resulted? Try to compute the molarity (moles/liter) of the final solution. Compare that to the pH measurement of it.

2. Reactants and Products: Magnesium (II) Oxide (a metal oxide), Water (a molecule), and Magnesium (II) Hydroxide (a strong base and an ionic compound) - MgO, H_2O, $Mg(OH)_2$. Chemical reaction equation: MgO + H_2O → $Mg(OH)_2$. Balanced Equation: MgO + H_2O → $Mg(OH)_2$.

MgO + H2O ▢ Mg(OH)2

1 mole of Magnesium Oxide=	40.31 Grams
1 mole of Water=	18.02 Grams

	MgO(magnesia, periclase)	H2O	Mg(OH)2(brucite)
Delta G			
Physical State	Col, cubic	Liquid	col, hex

Materials:

Magnesium (II) Oxide (a metal oxide) - MgO

Distilled Water (a molecule) - H_2O

Two 150 milliliter beakers

25 milliliter graduated cylinder

Weighing glassine paper

Digital or triple beam balance

Celsius thermometer

pH Meter

Glass rod

Procedure:

Tare a piece of glassine weighing paper on a balance. Using the balance, weigh out 4 grams of magnesium (II) oxide on the paper. Label one of the empty (clean, dry) 150ml beakers with the characters Mg) and weigh it on the balance. Record this as the empty MgO beaker mass. Transfer the 4 grams of magnesium oxide into it and weigh it again. Record this as the filled MgO beaker mass. Pour 20 milliliters of distilled water into the 25ml graduate cylinder (be sure to measure from the bottom of the meniscus). Weigh the other empty (clean, dry) 150ml beaker. Record this as the empty water beaker mass. Decant (pour) the 10ml of distilled water into it and weigh it again. Record this as the filled water beaker mass. Measure the temperature of the distilled water with the Celsius thermometer (call it the starting temperature). Measure the pH of the distilled water with the pH meter (call it the starting pH). Be prepared to observe all changes to the magnesium (II) oxide when you add the 20ml of distilled water to it. Decant (pour) the 20ml of distilled water from its 150ml beaker into the 150ml beaker containing the magnesium (II) oxide. Stir the solution with the glass rod. Visually observe the changes that occur in the resulting solution. Measure the resulting solution's temperature (call it the final temperature) and its pH. Measure the resulting solution's volume in the 25ml graduated cylinder.

Discussion and Conclusion Questions:

Write the balanced equation of the reaction that you have carried out: $MgO + H_2O \rightarrow Mg(OH)_2$. What type of reaction is this (combination, decomposition, substitution)? Does the temperature change indicate that an exothermic (heat was given off to the environment) or endothermic (heat was absorbed from the environment) reaction had occurred? Is the distilled water an acid, a base or neutral? Judging by the final pH, is the product an acid or base? From the mass of the distilled water, what is the number of moles of it being used in this chemical reaction? From the mass of the magnesium (II) oxide, how many moles of it are being used in this chemical reaction? From the number of moles of the two reactants (the water and the magnesium oxide), how many moles of magnesium (II) hydroxide theoretically could result (assuming a 100% yield)? From the mass of the final solution what % yield of $Mg(OH)_2$ appears to have resulted? Try to compute the molarity (moles/liter) of the final solution. Compare that to the pH measurement of it.

B. A Combination Reaction of a Nonmetal Oxide with Water that Produces an Oxyacid in which the Nonmetal is in the same Oxidation State as in the starting Oxide:

1. Reactants and Products: Sulfur Dioxide (a non-metal oxide), Water (a molecule), and Sulfurous Acid (an oxyacid) - SO_2, H_2O, H_2SO_3. Chemical reaction equation:

$SO_2 + H_2O \rightarrow H_2SO_3$. Balanced Equation: $SO_2 + H_2O \rightarrow H_2SO_3$.

SO2 + H2O ☐ H2SO3

| 1 mole of Sulfur Dioxide= | 64.07 Grams |
| 1 mole of Water= | 18.02 Grams |

	SO2	H2O	H2SO3
Delta G			
Physical State	col, gas	Liquid	in sol only

Materials:

Sulfur Dioxide (a metal oxide) - SO_2

Distilled Water (a molecule) - H_2O

Two 150 milliliter beakers

25 milliliter graduated cylinder

Weighing glassine paper

Digital or triple beam balance

Celsius thermometer

pH meter

Glass rod

Procedure:

Tare a piece of glassine weighing paper on a balance. Using the balance, weigh out 6 grams of sulfur dioxide on the paper. Label one of the empty (clean, dry) 150ml beakers with the characters SO₂ and weigh it on the balance. Transfer the 6 grams of sulfur dioxide into it and weigh it again. Pour 20 milliliters of distilled water into the 25ml graduate cylinder (be sure to measure from the bottom of the meniscus). Weigh the other empty (clean, dry) 150ml beaker. Record this as the empty water beaker mass. Decant (pour) the 20ml of distilled water into it and weigh it again. Record this as the filled water beaker mass. Measure the temperature of the distilled water with the Celsius thermometer (call it the starting temperature). Measure the pH of the distilled water with the pH meter (call it the starting pH). Be prepared to observe all changes to the sulfur dioxide when you add the 20ml of distilled water to it. Decant (pour) the 20ml of distilled water to the sulfur dioxide. Stir the resulting solution with the glass rod. Visually observe the changes that occur in the resulting solution. Measure the resulting solution's temperature (call it the final temperature) and its pH. Measure the resulting solution's volume in the 25ml graduated cylinder.

Discussion and Conclusion Questions:

Write the balanced equation of the reaction that you have carried out: $SO_2 + H_2O \rightarrow H_2SO_3$. What type of reaction is this (combination, decomposition, substitution)? Does the temperature change indicate that an exothermic (heat was given off to the environment) or endothermic (heat was absorbed from the environment) reaction had occurred? Is the distilled water an acid, a base or neutral? Judging by the final pH is the product an acid or base? From the mass of the distilled water, what is the number of moles of it being used in this chemical reaction? From the mass of the sulfur oxide, how many moles of it are being used in this chemical reaction? From the number of moles of the two reactants (the water and the sulfur oxide), how many moles of sulfurous acid theoretically could result (assuming a 100% yield)? From the mass of the final solution what % yield of H_2SO_3 appears to have resulted? Try to compute the molarity (moles/liter) of the final solution. Compare that to the pH measurement of it.

C. Decomposition Reaction (actually a reversal of a combination reaction accomplished by heating the product of the combination reaction between a metal oxide and water):

1. Reactants and Products: Iron (III) Hydroxide (a metal hydroxide), Iron (III) Oxide (a metal oxide), and Water (a molecule) - $Fe(OH)_3$, Fe_2O_3, H_2O. Chemical reaction equation:

$Fe(OH)_3 \rightarrow Fe_2O_3 + H_2O$. Balanced Equation: $2Fe(OH)_3 \rightarrow Fe_2O_3 + 3H_2O$.

$2Fe(OH)3 \square Fe2O3 + 3H2O$

2 moles of Iron (II) Hydroxide=   179.74 Grams

	2Fe(OH)3	Fe2O3(hematite)	3H2O
Delta G			
Physical State	rusty colored	red-brn, trig	Liquid

Materials:

Iron (III) Hydroxide (a metal hydroxide) - $Fe(OH)_3$

Small porcelain-weighing dish

Weighing glassine paper

Digital or triple beam balance

Ring stand

Wire mesh square with heat resistant material at its center

Bunsen burner

Striker

Procedure:

Tare a piece of glassine weighing paper on a balance. Using the balance, weigh out 9 grams of iron (III) hydroxide on the paper. Transfer the out 9 grams of iron (III) hydroxide into the Small porcelain-weighing dish. Be prepared to observe all changes to the iron (III) hydroxide when you heat it on the ring stand with the Bunsen burner. Set up the Bunsen burner under the ring attached firmly to the ring stand. Place the wire mesh square on the ring and place the small porcelain-weighing dish containing the iron (III) hydroxide. Observe the changes that occur in the small porcelain-weighing dish as you heat it.

D. Combination Reactions of Metal Oxides and Nonmetal Oxides to give Oxysalts:

1. Reactants and Products: Calcium Oxide (a metal oxide), Sulfur Trioxide (a non-metal oxide), and Calcium Sulfate (an Oxysalt) - only if water is absent - CaO, SO_3, $CaSO_4$(s). Chemical reaction equation: $CaO + SO_3 \rightarrow CaSO_4$. Balanced Equation: $CaO + SO_3 \rightarrow CaSO_4$.

CaO + SO3 ☐ CaSO4

1 mole of Calcium Oxide=		56.08 Grams
1 mole of Sulfur Trioxide=		80.07 Grams
	CaO	SO3	CaSO4(anhydrite(
Delta G			
Physical State	col, cubic	asbestos-like nd	col, rh.

2. Reactants and Products: Sodium Hydroxide (a metal oxide), Carbon Dioxide (a non-metal oxide), and Sodium Bicarbonate (a Oxysalt) - only if water is not present - NaOH, CO_2, $NaHCO_3$. Chemical reaction equation: $NaOH + CO_2 \rightarrow NaHCO_3$. Balanced Equation: $NaOH + CO_2 \rightarrow NaHCO_3$.

NaOH + CO2 ☐ NaHCO3

1 mole of Sodium Hydroxide=		40 Grams
1 mole of Carbon Dioxide=		44.01 Grams
	NaOH	CO2	NaHCO3(sodium hydrogen carbonate)
Delta G			
Physical State	white, deliq	Gas	Wh pwd, mn

Materials:

10 Molar Sodium Hydroxide (a metal oxide) – 10M NaOH(aq)

Frozen Carbon Dioxide (a non-metal oxide) – dry ice - CO_2(s)

150 milliliter beaker

50 milliliter graduated cylinder

Digital balance or triple-beam balance

Celsius thermometer

pH meter

Glass stirring rod

Procedure:

Pour 20 milliliters of 10 Molar sodium hydroxide into the 50ml graduated cylinder (be sure to measure from the bottom of the meniscus). Label the empty (clean, dry) 150ml beaker with the characters NaOH and weigh it on the balance. Record this as the empty mass of the NaOH beaker. Transfer the 20 milliliters of 10M sodium hydroxide into it and weigh it again. Record this as the filled NaOH beaker mass. Measure its pH using the pH meter. Record this as its starting pH. Be prepared to observe all changes to the sodium hydroxide solution when you add a chip of dry ice to it. Add a chip of dry ice (solid carbon dioxide) to the sodium hydroxide solution (which is now in the 150ml beaker). Stir the resulting mixture with the glass rod. Visually observe the changes that occur in the resulting solution. Measure the resulting solution's temperature (call it the final temperature) and its pH. Measure the resulting solution's volume in the 50ml graduated cylinder.

Discussion and Conclusion Questions:

Write the balanced equation of the reaction that you have carried out: $NaOH + CO_2 \rightarrow NaHCO_3$. What type of reaction is this (combination, decomposition, substitution)? Does the temperature change indicate that an exothermic (heat was given off to the environment) or endothermic (heat was absorbed from the environment) reaction had occurred? Is the sodium hydroxide an acid, a base or neutral? Judging by the final pH is the product an acid or base? From the mass of the sodium hydroxide, what is the number of moles of it being used in this chemical reaction? Compute the number of moles of sodium hydroxide from its volume, 20ml, its molarity, 10 moles/liter. How do these two figures compare? From the number of moles of the sodium hydroxide, how many moles of $NaHCO_3$ (sodium hydrogen carbonate or sodium bicarbonate) theoretically could result (assuming a 100% yield)? From the mass of the final solution what % yield of $NaHCO_3$ appears to have resulted? Try to compute the molarity (moles/liter) of the final solution. Compare that to the pH measurement of it.

E. Acid-Base Neutralization Reactions where an Acid reacts with a Base to form a Salt and Water:

1. Reactants and Products: Hydrochloric Acid (a strong acid), Calcium Hydroxide (a strong base), Calcium Chloride (a metal salt), Water (a molecule) - HCl, $Ca(OH)_2$, $CaCl_2$, H_2O. Chemical reaction equation: $HCl + Ca(OH)_2 \rightarrow CaCl_2 + H_2O$. Balanced Equation: $2HCl + Ca(OH)_2 \rightarrow CaCl_2 + 2H_2O$.

$2HCl + Ca(OH)2 \rightarrow CaCl2 + 2H2O$

2 moles of Hydrochloric Acid=   36.46 Grams

1 mole of Calcium Hydroxide=   74.1 Grams

	2HCl	Ca(OH)2	CaCl2	2H2O
Delta G		-62.74		
Physical State	Liquid	col, hex	col, cub	Liquid

Materials:

10 Molar Hydrochloric Acid (a strong acid) – 10M HCl

10 Molar Calcium Hydroxide (a strong base) – 10M $Ca(OH)_2$,

Three 150 milliliter beakers

50 milliliter graduated cylinder

Digital or triple beam balance

Celsius thermometer

pH meter

Glass stirring rod

Procedure:

Decant (pour) about 25 milliliters of 10 Molar hydrochloric acid from the stock bottle into one of the empty (clean, dry) 150ml beakers. Do not pour back into the stock bottle. Decant (pour) 20 milliliters of HCl into the 50ml graduate cylinder (be sure to measure from the bottom of the meniscus). Wash the 150ml beaker thoroughly with tap water. Label with the letters HCl and then weigh another empty (clean, dry) 150ml beaker on the balance. Record the mass as the empty HCl beaker mass. Transfer the 20 milliliters of 10M hydrochloric acid into it and weigh it again. Record this as the filled HCL beaker mass. Measure the HCl's temperature with the Celsius thermometer. Record this as the starting HCL temperature. After removing the thermometer from the beaker, wash the thermometer off with tap water. Measure the HCl's pH using the pH meter. Record this as the HCl pH. After removing the pH electrode from the beaker, thoroughly rinse it off with distilled water. Try to avoid getting HCl on you. Rinse the 50ml graduated cylinder thoroughly with tap water. Decant (pour) about 15 milliliters of the 10 Molar calcium hydroxide solution into the just washed 150ml beaker from the stock bottle. Do not pour back into the stock bottle. Decant (pour) 10ml of $Ca(OH)_2$ into the 50ml graduated cylinder (be sure to measure from the bottom of the meniscus). Label the remaining empty (clean, dry) 150 beaker with the characters $Ca(OH)_2$ and weigh it. Record this mass as the empty $Ca(OH)_2$ beaker mass. Decant the 10ml of $Ca(OH)_2$ into the beaker and weigh it. Record this as the filled $Ca(OH)_2$ beaker mass. Measure the temperature of the calcium hydroxide with the Celsius thermometer. Record this as the $Ca(OH)_2$ temperature. After removing the thermometer from the beaker, wash the thermometer thoroughly with tap water. Measure the pH of the calcium hydroxide with the pH meter. Record this as the $Ca(OH)_2$ pH. After removing the pH electrode from the beaker, wash it off thoroughly with distilled water. Be prepared to observe all changes to the 10M hydrochloric acid solution when you slowly add the 10M calcium hydroxide solution to it. Slowly add the calcium hydroxide to the hydrochloric acid. Stir the solution with the glass rod. Visually observe the changes that occur in the resulting solution. Measure the resulting solution's temperature (call it the final temperature) and its pH and weigh it (record it as the calcium chloride solution mass). Leave the beaker (which you had weighed previously) out to evaporate the remaining water. Weigh the dry beaker and record its mass as the calcium chloride mass.

Discussion and Conclusion Questions:

Write the balanced equation of the reaction that you have carried out: HCl + $Ca(OH)_2$ → $CaCl_2$ + H_2O. What type of reaction is this (combination, decomposition, substitution, acid-base neutralization)? Does the temperature change indicate that an exothermic (heat was given off to the environment) or endothermic (heat was absorbed from the environment) reaction had occurred? Judging by it pH is the HCl an acid, a base or neutral? Judging by its pH is the calcium hydroxide an acid, a base, or neutral? Judging by the final pH, is the product mixture an acid or base or neutral? From the mass of the hydrochloric acid, what is the number of moles of it being used in this chemical reaction? Calculate the number of moles of HCl present at the start of the reaction from the volume used in milliliters and its molarity (moles/liter) of 10 Molar. From the mass of the calcium hydroxide, how many moles of it are present at the start of this chemical reaction? Calculate the number of moles of it at the start of the reaction from its volume in milliliters and its molarity (moles/liter) of 10 Molar. From the number of moles of the two reactants (the hydrochloric acid and the calcium hydroxide), how many moles of calcium chloride theoretically could result (assuming a 100% yield)? From the pH of the final solution, is it an acid, base, or neutral? From the mass of the dry beaker with the calcium chloride crstals, what % yield of $CaCl_2$ appears to have resulted?

2. Reactants and Products: Sulfuric Acid (a strong acid), Iron (III) Hydroxide (a strong base), Iron (III) Sulfate (a metal salt), and Water (a molecule) - H_2SO_4, $Fe(OH)_3$, $Fe_2(SO_4)_3$, H_2O. Chemical reaction equation: $H_2SO_4 + Fe(OH)_3 \rightarrow Fe_2(SO_4)_3 + H_2O$. Balanced Equation: $3H_2SO_4 + 2Fe(OH)_3 \rightarrow Fe_2(SO_4)_3 + 6H_2O$.

3H2SO4 + 2Fe(OH)3 ▢ Fe2(SO4)3 + 6H2O

3 moles of Sulfuric Acid=   294.27 Grams

2 moles of Iron (II) Hydroxide=   213.76 Grams

	3H2SO4	2Fe(OH)3	Fe2(SO4)3 (iron(III) sulfate)	6H2O
Delta G	-494.79	-232.6	-540.9	-226.748
Physical State	Liquid	rust colored	yel, rh	liquid

Materials:

10 Molar Sulfuric Acid (a strong acid) – 10M H_2SO_4

10 Molar Iron (III) Hydroxide (a strong base) – 10M $Fe(OH)_3$

Three 150 milliliter beakers

50 milliliter graduated cylinder

Weighing glassine paper

Digital or triple beam balance

Celsius thermometer

pH meter

Glass stirring rod

Procedure:

Decant (pour) about 20 milliliters of 10 Molar sulfuric acid from the stock bottle into one of the empty (clean, dry) 150ml beakers. Do not pour back into the stock bottle. Decant (pour) 15 milliliters of sulfuric acid into the 50ml graduate cylinder (be sure to measure from the bottom of the meniscus). Wash the 150ml beaker thoroughly with tap water. Label with the letters H_2SO_4 and then weigh another empty (clean, dry) 150ml beaker on the balance. Record the mass as empty H₂SO₄ beaker mass. Transfer the 15 milliliters of 10M sulfuric acid into it and weigh it again. Record this as the filled H_2SO_4 beaker mass. Measure the H_2SO_4's temperature with the Celsius thermometer. Record this as the starting H_2SO_4 temperature. After removing the thermometer from the beaker, wash the thermometer off with tap water. Measure the H_2SO_4's pH using the pH meter. Record this as the H_2SO_4 pH. After removing the pH electrode from the beaker, thoroughly rinse it off with distilled water. Try to avoid getting H_2SO_4 on you. Rinse the 50ml graduated cylinder thoroughly with tap water. Decant (pour) about 15 milliliters of the 10 Molar iron (III) hydroxide into the just washed 150ml beaker from the stock bottle. Do not pour back into the stock bottle. Decant (pour) 10ml of $Fe(OH)_3$ into the 50ml graduated cylinder (be sure to measure from the bottom of the meniscus). Label the remaining empty (clean, dry) 150 beaker with the characters $Fe(OH)_3$ and weigh it. Record this mass as the empty iron (III) hydroxide mass. Decant the 10ml of iron hydroxide into this beaker and weigh it. Record this as the filled $Fe(OH)_3$ beaker mass. Measure the temperature of the iron (III) hydroxide with the Celsius thermometer. Record this as the $Fe(OH)_3$ temperature. After removing the thermometer from the beaker, wash the thermometer thoroughly with tap water. Measure the pH of the iron (III) hydroxide with the pH meter. Record this as the $Fe(OH)_3$ pH. After removing the pH electrode from the beaker, wash it off thoroughly with distilled water. Be prepared to observe all changes to the 10M sulfuric acid solution when you slowly add the 10M iron (III) hydroxide solution to it. Slowly add the iron (III) hydroxide to the sulfuric acid. Stir the solution with the glass rod. Visually observe the changes that occur in the resulting solution. Measure the resulting solution's temperature (call it the final temperature) and its pH. Leave the beaker (which you had weighed previously) out to evaporate the remaining water. Weigh the dry beaker and record it's mass as the iron (III) sulfate mass.

Discussion and Conclusion Questions:

Write the balanced equation of the reaction that you have carried out: $H_2SO_4 + Fe(OH)_3 \rightarrow Fe_2(SO_4)_3 + H_2O$. What type of reaction is this (combination, decomposition, substitution, acid-base neutralization)? Does the temperature change indicate that an exothermic (heat was given off to the environment) or endothermic (heat was absorbed from the environment) reaction had occurred? Judging by it pH is the H_2SO_4 an acid, a base or neutral? Judging by its pH is the iron (III) hydroxide an acid, a base, or neutral? Judging by the final pH, is the product mixture an acid or base or neutral? From the mass of the sulfuric acid, what is the number of moles of it being used in this chemical reaction? Calculate the number of moles of H_2SO_4 present at the start of the reaction from the volume used in milliliters and its molarity (moles/liter) of 10 Molar. From the mass of the $Fe(OH)_3$, how many moles of it are present at the start of this chemical reaction? Calculate the number of moles of it at the start of the reaction from its volume in milliliters and its molarity (moles/liter) of 10 Molar. From the number of moles of the two reactants (the sulfuric acid and the iron (III) hydroxide), how many moles of iron (III) sulfate theoretically could result (assuming a 100% yield)? From the weight of the final solution what % yield of $Fe_2(SO_4)_3$ appears to have resulted? From the pH of the final solution, is it an acid, base, or neutral?

3. Reactants and Products: Aluminum Oxide (a weak base), Hydrochlorous Acid (a weak acid), Aluminum Chlorate (a metal salt), and Water (a molecule) - Al_2O_3, $HClO_4$, $Al(ClO_4)_3$, H_2O. Chemical reaction equation: $Al_2O_3 + HClO_4 \rightarrow Al(ClO_4)_3 + H_2O$. Balanced Equation: $Al_2O_3 + 6HClO_4 \rightarrow 2Al(ClO_4)_3 + 3H_2O$.

Al2O3 + 6HclO4 ☐ 2Al(ClO4)3 + 3H2O

1 mole of Aluminum Oxide+   101.96 grams

6 moles of Hydrochlorous Acid=   602.76 grams

	Al2O3	6HClO4	2Al(ClO4)3	3H2O
Delta G			hrdrate	
Physical State	wh micr cr	liquid	col,rhhd,deliq	liquid

Materials:

Aluminum Oxide (a weak base) - Al_2O_3

10 Molar Hydrochlorous Acid (a weak acid) – 10M $HClO_4$,

Three 150 milliliter beakers

100 milliliter graduated cylinder

Weighing glassine paper

Digital or triple beam balance

Celsius thermometer

pH meter

Glass stirring rod

Procedure:

Tare a piece of glassine weighing paper on a balance. Using the balance, weigh out 5 grams of aluminum oxide on the paper. Label one of the empty (clean, dry) 150ml beakers with the characters Al_2O_3 and weigh it on the balance. Transfer the 5 grams of aluminum oxide into it and weigh it again. Record the mass as filled Al_2O_3 beaker mass. Decant (pour) about 40 milliliters of the 10 Molar iron (III) hydroxide into and empty (clean, dry) 150ml beaker from the stock bottle. Do not pour back into the stock bottle. Decant (pour) 30ml of $HClO_4$ into the 100ml graduated cylinder (be sure to measure from the bottom of the meniscus). Label the remaining empty (clean, dry) 150 beaker with the characters $HClO_4$ and weigh it. Record this mass as the empty $HClO_4$ beaker mass. Decant the 30ml of $HClO_4$ from the 100ml graduated cylinder into this beaker and reweigh it. Record this as the filled $HClO_4$ beaker mass. Measure the temperature of the hydrochlorous acid with the Celsius thermometer. Record this as the $HClO_4$ temperature. After removing the thermometer from the beaker, wash the thermometer thoroughly with tap water. Measure the pH of the hydrochlorous acid with the pH meter. Record this as the $HClO_4$ pH. After removing the pH electrode from the beaker, wash it off thoroughly with distilled water. Be prepared to observe all changes to the Al_2O_3 when you slowly add the 10M hydrochlorous acid to it. Slowly add the hydrochlorous acid to the aluminum oxide. Stir the solution with the glass rod. Visually observe the changes that occur in the resulting solution. Measure the resulting solution's temperature (call it the final temperature) and its pH. Leave the beaker (which you had weighed previously) out to evaporate the remaining water. Weigh this beaker when dry and record its mass as the aluminum chlorate beaker mass.

Discussion and Conclusion Questions:

Write the balanced equation of the reaction that you have carried out: Al_2O_3 + $HClO_4$ → $Al(ClO_4)_3$ + H_2O. What type of reaction is this (combination, decomposition, substitution, acid-base neutralization)? Does the temperature change indicate that an exothermic (heat was given off to the environment) or endothermic (heat was absorbed from the environment) reaction had occurred? Judging by its pH is the $HClO_4$ an acid, a base, or neutral? Judging by the final pH, is the product mixture an acid or base or neutral? From the mass of the aluminum oxide, what is the number of moles of it at the start of this chemical reaction? From the mass of the $HClO_4$, how many moles of it are present at the start of this chemical reaction? Calculate the number of moles of it at the start of the reaction from its volume in milliliters and its molarity (moles/liter) of 10 Molar. From the number of moles of the two reactants (the aluminum oxide and the hydrochlorous acid), how many moles of aluminum chlorate theoretically could result (assuming a 100% yield)? From the mass of the final product what % yield of $Al(ClO_4)_3$ appears to have resulted?

F. An Acid-Base Reaction where Ammonium Salts react with Metal Oxides to produce Ammonia:

1. Reactants and Products: Ammonium Nitrate (an ionic compound), Calcium Oxide (a metal oxide), Ammonia, Water, and Calcium Nitrate - NH_4NO_3, CaO, NH_3, H_2O, $Ca(NO_3)_2$. Chemical reaction equation: NH_4NO_3, CaO, $NH_4NO_3 + CaO \rightarrow NH_3 + H_2O + Ca(NO_3)_2$. Balanced Equation: $2NH_4NO_3 + CaO \rightarrow 2NH_3 + H_2O + Ca(NO_3)_2$.

2NH4NO3 + CaO ☐ 2NH3 + H2O + Ca(NO3)2

2 moles of Ammonium Nitrate=	160.12 grams	
1 mole of Calcium Oxide=	56.08 grams	

	2NH4NO3	CaO	2NH3	H2O	Ca(NO3)2
Delta G				-56.687	
Physical State	col, rh	col, cubic	gas	liquid	col, cub, hygr

G. Acid-Base Reactions in which the Salt of a Weak Acid (contains the anion of a weak acid) reacts with a Strong Acid to make a Weak Acid and a Salt:

1. Reactants and Products: Sodium Bicarbonate (salt of a weak acid), Sulfuric Acid (strong acid), Sodium Sulfate (oxyacid), Carbon Dioxide (molecule), and Water - $NaHCO_3$, H_2SO_4, Na_2SO_4, CO_2, H_2O. Chemical reaction equation: $NaHCO_3 + H_2SO_4 \rightarrow Na_2SO_4 + CO_2 + H_2O$. Balanced Equation:

$2NaHCO_3 + H_2SO_4 \rightarrow Na_2SO_4 + 2CO_2 + 2H_2O$.

2NaHCO3 + H2SO4 ☐ Na2SO4 + 2CO2 + 2H2O

2 moles of Sodium Bicarbonate=   168.02 grams

1 mole of Sulfuric Acid=   98.09 grams

	2NaHCO3	H2SO4	Na2SO4 (mirailite, thenardite)	2CO2	2H2O
Delta G					
Physical State	col, rh (d35)	liquid	col, rh-bipyr	Gas	liquid

Materials:

Sodium Bicarbonate (salt of a weak acid) - $NaHCO_3$

10 Molar Sulfuric Acid (strong acid) – 10M H_2SO_4

Three 150 milliliter beakers

50 milliliter graduated cylinder

Weighing glassine paper

Digital or triple beam balance

Celsius thermometer

pH meter

Glass stirring rod

Procedure:

Tare a piece of glassine weighing paper on a balance. Using the balance, weigh out 17 grams of sodium bicarbonate on the paper. Label one of the empty (clean, dry) 150ml beakers with the characters $NaHCO_3$ and weigh it on the balance. Record this as the empty $NaHCO_3$ beaker mass. Transfer the 17 grams of sodium bicarbonate into it and weigh it again. Record the mass as filled $NaHCO_3$ beaker mass. Decant (pour) about 60 milliliters of the 10 Molar sulfuric acid into another empty (dry, clean) 150ml beaker from the stock bottle. Do not pour back into the stock bottle. Decant (pour) 50ml of H_2SO_4 into the 50ml graduated cylinder (be sure to measure from the bottom of the meniscus). Label the remaining empty (clean, dry) 150 beaker with the characters H_2SO_4 and weigh it. Record this mass as the empty H_2SO_4 beaker mass. Decant the 50ml of H_2SO_4 from the graduated cylinder into this beaker and reweigh it. Record this as the filled H_2SO_4 beaker mass. Measure the temperature of the sulfuric acid with the Celsius thermometer. Record this as the H_2SO_4 temperature. After removing the thermometer from the beaker, wash the thermometer thoroughly with tap water. Measure the pH of the sulfuric acid with the pH meter. Record this as the H_2SO_4 pH. After removing the pH electrode from the beaker, wash it off thoroughly with distilled water. Be prepared to observe all changes to the $NaHCO_3$ when you slowly add the 10M sulfuric acid to it. Slowly add the sulfuric acid to the aluminum oxide. Stir the solution with the glass rod. Visually observe the changes that occur in the resulting solution. Measure the resulting solution's temperature (call it the final temperature) and its pH. . Leave the beaker (which you had weighed previously) out to evaporate the remaining water. Weigh this beaker hen dry and record its mass as the sodium sulfate beaker mass.

Discussion and Conclusion Questions:

Write the balanced equation of the reaction that you have carried out: $NaHCO_3 + H_2SO_4 \rightarrow Na_2SO_4 + CO_2 + H_2O$. What type of reaction is this (combination, decomposition, substitution, acid-base neutralization)? Does the temperature change indicate that an exothermic (heat was given off to the environment) or endothermic (heat was absorbed from the environment) reaction had occurred? Judging by its pH is the H_2SO_4 an acid, a base, or neutral? Judging by the final pH, is the product mixture an acid or base or neutral? From the mass of the sodium bicarbonate, what is the number of moles of it at the start of this chemical reaction? From the mass of the H_2SO_4, how many moles of it are present at the start of this chemical reaction? Calculate the number of moles of it at the start of the reaction from its volume in milliliters and its molarity (moles/liter) of 10 Molar. From the number of moles of the two reactants (the sodium bicarbonate and the sulfuric acid), how many moles of sodium sulfate theoretically could result (assuming a 100% yield)? From the weight of the final product what % yield of Na_2SO_4 appears to have resulted?

2. Reactants and Products: Barium Carbonate (salt of a weak acid), Hydrobromic Acid (weak acid), Barium DiBromide (salt), Carbon Dioxide (molecule) and Water - $BaCO_3$, HBr, $BaBr_2$, CO_2, H_2O. Chemical reaction equation: $BaCO_3 + HBr \rightarrow BaBr_2 + CO_2 + H_2O$.

Balanced Equation: $BaCO_3 + 2HBr \rightarrow BaBr_2 + CO_2 + H_2O$.

BaCO3 + 2HBr ☐ BaBr2 + CO2 + H2O

| 1 mole of Barium Carbonate= | 197.35 grams |
| 2 moles of Hydrobromic Acid= | 161.82 grams |

	BaCO3	2HBr	BaBr2	CO2	H2O
Delta G					
Physical State	wh hex	liquid	col cr	gas	liquid

Materials:

Barium Carbonate (salt of a weak acid) – $BaCO_3$

10 Molar Hydrobromic Acid (weak acid) – 10M HBr

Three 150 milliliter beakers

25 milliliter graduated cylinder

Weighing glassine paper

Digital or triple beam balance

Celsius thermometer

pH meter

Glass stirring rod

Procedure:

Tare a piece of glassine weighing paper on a balance. Using the balance, weigh out 20 grams of barium carbonate on the paper. Label one of the empty (clean, dry) 150ml beakers with the characters $BaCO_3$ and weigh it on the balance. Record this as the empty $BaCO_3$ beaker mass. Transfer the 20 grams of barium carbonate into it and weigh it again. Record the mass as filled $BaCO_3$ beaker mass. Decant (pour) about 15 milliliters of the 10 Molar hydrobromic acid into another empty (dry, clean) 150ml beaker from the stock bottle. Do not pour back into the stock bottle. Decant (pour) 10ml of HBr into the 25ml graduated cylinder (be sure to measure from the bottom of the meniscus). Label the remaining empty (clean, dry) 150 beaker with the characters HBr and weigh it. Record this mass as the empty HBr beaker mass. Decant the 10ml of HBr from the graduated cylinder into this beaker and reweigh it. Record this as the filled HBr beaker mass. Measure the temperature of the hydrobromic acid with the Celsius thermometer. Record this as the HBr temperature. After removing the thermometer from the beaker, wash the thermometer thoroughly with tap water. Measure the pH of the hydrobromic acid with the pH meter. Record this as the HBr. After removing the pH electrode from the beaker, wash it off thoroughly with distilled water. Be prepared to observe all changes to the $BaCO_3$ when you slowly add the 10M HBr to it. Slowly add the hydrobromic acid to the barium carbonate. Stir the solution with the glass rod. Visually observe the changes that occur in the resulting solution. Measure the resulting solution's temperature (call it the final temperature) and its pH. Leave the beaker (which you had weighed previously) out to evaporate the remaining water. Re-weigh this beaker and record it as the $BaBr_2$ beaker mass.

Discussion and Conclusion Questions:

Write the balanced equation of the reaction that you have carried out: $BaCO_3 + HBr \rightarrow BaBr_2 + CO_2 + H_2O$. What type of reaction is this (combination, decomposition, substitution, acid-base neutralization)? Does the temperature change indicate that an exothermic (heat was given off to the environment) or endothermic (heat was absorbed from the environment) reaction had occurred? Judging by its pH is the HBr an acid, a base, or neutral? Judging by the final pH, is the product mixture an acid or base or neutral? From the mass of the barium carbonate, what is the number of moles of it at the start of this chemical reaction? From the mass of the HBr how many moles of it are present at the start of this chemical reaction? Calculate the number of moles of it at the start of the reaction from its volume in milliliters and its molarity (moles/liter) of 10 Molar. From the number of moles of the two reactants (the barium carbonate and the hyrdrobromic acid), how many moles of barium bromide theoretically could result (assuming a 100% yield)? From the mass of the final product what % yield of $BaBr_2$ appears to have resulted?

3. Reactants and Products: Magnesium Sulfide (salt of a weak acid), Hydrochloric Acid (strong acid), Hydrogen Sulfide (weak acid), and Magnesium Chloride (salt) - MgS, HCl, H_2S, $MgCl_2$. Chemical reaction equation: MgS + HCl → H_2S + $MgCl_2$. Balanced Equation: MgS + 2HCl → H_2S + $MgCl_2$.

MgS + 2HCl ☐ H2S + MgCl2

1 moles of MagnesiumSulfide=   56.38 grams

2 moles of Hydrochloric Acid=   72.92 grams

	MgS	2HCl	H2S	MgCl2
Delta G				
Physical State		liquid	Gas	col, hex

Materials:

Magnesium Sulfide (salt of a weak acid) – MgS

10 Molar Hydrochloric Acid (strong acid) – 10M HCl,

Three 150 milliliter beakers

50 milliliter graduated cylinder

Weighing glassine paper

Digital or triple beam balance

Celsius thermometer

pH meter

Glass stirring rod

Procedure:

Tare a piece of glassine weighing paper on a balance. Using the balance, weigh out 6 grams of magnesium sulfide on the paper. Label one of the empty (clean, dry) 150ml beakers with the characters MgS and weigh it on the balance. Record this as the empty MgS beaker mass. Transfer the 6 grams of magnesium sulfide into it and weigh it again. Record the mass as filled MgS beaker mass. Decant (pour) about 25 milliliters of the 10 Molar hydrochloric acid into another empty (dry, clean) 150ml beaker from the stock bottle. Do not pour back into the stock bottle. Decant (pour) 20ml of HCl into the 50ml graduated cylinder (be sure to measure from the bottom of the meniscus). Label the remaining empty (clean, dry) 150 beaker with the characters HCl and weigh it. Record this mass as the filled HCl beaker mass. Measure the temperature of the hydrochloric acid with the Celsius thermometer. Record this as the HCl temperature. After removing the thermometer from the beaker, wash the thermometer thoroughly with tap water. Measure the pH of the hydrochloric acid with the pH meter. Record this as the HCl. After removing the pH electrode from the beaker, wash it off thoroughly with distilled water. Be prepared to observe all changes to the MgS when you slowly add the 10M HCl to it. Slowly add the hydrochloric acid to the magnesium sulfide. Stir the solution with the glass rod. Visually observe the changes that occur in the resulting solution. Measure the resulting solution's temperature (call it the final temperature) and its pH. Leave the beaker (which you had weighed previously) out to evaporate the remaining water. Re-weigh this beaker and record it as the $MgCl_2$ beaker mass.

Discussion and Conclusion Questions:

Write the balanced equation of the reaction that you have carried out: $MgS + HCl \rightarrow H_2S + MgCl_2$ What type of reaction is this (combination, decomposition, substitution, acid-base neutralization)? Does the temperature change indicate that an exothermic (heat was given off to the environment) or endothermic (heat was absorbed from the environment) reaction had occurred? Judging by its pH is the HCl an acid, a base, or neutral? Judging by the final pH, is the product mixture an acid or base or neutral? From the mass of the magnesium sulfide, what is the number of moles of it at the start of this chemical reaction? From the mass of the HCl how many moles of it are present at the start of this chemical reaction? Calculate the number of moles of it at the start of the reaction from its volume in milliliters and its molarity (moles/liter) of 10 Molar. From the number of moles of the two reactants (the magnesium sulfide and the hydrochloric acid), how many moles of magnesium chloride theoretically could result (assuming a 100% yield)? From the mass of the final product what % yield of $MgCl_2$ appears to have resulted?

4. Reactants and Products: Potassium Sulfite (salt of a strong acid), Nitric Acid (strong acid), Potassium Nitrate (weak acid), Sulfur Dioxide (molecule), and Water - K_2SO_3, HNO_3, KNO_3, SO_2, H_2O. Chemical reaction equation: $K_2SO_3 + HNO_3 \rightarrow KNO_3 + SO_2 + H_2O$.

Balanced Equation: $K_2SO_3 + 2HNO_3 \rightarrow 2KNO_3 + SO_2 + H_2O$.

K2SO3 + 2HNO3 ▢ 2KNO3 + SO2 + H2O

1 mole of Potassium Sulfite= 158.27 grams

2 moles of Nitic Acid= 126.04 grams

	K2SO3(sufite-dihydrate)	2HNO3	2KNO3(saltpeter)	SO2	H2O
Delta G					
Physical State	w,cr,rh	liquid	col, rh	gas	liquid

Materials:

Potassium Sulfite (salt of a strong acid) - K_2SO_3

10 Molar Nitric Acid (strong acid) – 10M HNO_3

Three 150 milliliter beakers

50 milliliter graduated cylinder

Weighing glassine paper

Digital or triple beam balance

Celsius thermometer

pH meter

Glass stirring rod

Procedure:

Tare a piece of glassine weighing paper on a balance. Using the balance, weigh out 16 grams of potassium sulfite on the paper. Label one of the empty (clean, dry) 150ml beakers with the letters K_2SO_3 and weigh it on the balance. Record this as the empty K_2SO_3 beaker mass. Transfer the 20 grams of potassium sulfite into it and weigh it again. Record the mass as filled K_2SO_3 beaker mass. Decant (pour) about 25 milliliters of the 10 Molar nitric acid into another empty (dry, clean) 150ml beaker from the stock bottle. Do not pour back into the stock bottle.

Decant (pour) 20ml of HNO_3 into the 50ml graduated cylinder (be sure to measure from the bottom of the meniscus). Label the remaining empty (clean, dry) 150 beaker with the characters HNO_3 and weigh it. Record this mass as the empty HNO_3 beaker mass. Decant the 20ml of HNO_3 from the graduated cylinder into this beaker. Record this as the filled HNO_3 beaker mass. Measure the temperature of the nitric acid with the Celsius thermometer. Record this as the HNO_3 temperature. After removing the thermometer from the beaker, wash the thermometer thoroughly with tap water. Measure the pH of the nitric acid with the pH meter. Record this as the HNO_3. After removing the pH electrode from the beaker, wash it off thoroughly with distilled water. Be prepared to observe all changes to the K_2SO_3 when you slowly add the 10M HNO_3 to it. Slowly add the nitric acid to the potassium sulfite. Stir the solution with the glass rod. Visually observe the changes that occur in the resulting solution. Measure the resulting solution's temperature (call it the final temperature) and its pH. Leave the beaker (which you had weighed previously) out to evaporate the remaining water. Re-weigh this beaker and record it as the KNO_3 beaker mass.

Discussion and Conclusion Questions:

Write the balanced equation of the reaction that you have carried out: $K_2SO_3 + HNO_3 \rightarrow KNO_3 + SO_2 + H_2O$. What type of reaction is this (combination, decomposition, substitution, acid-base neutralization)? Does the temperature change indicate that an exothermic (heat was given off to the environment) or endothermic (heat was absorbed from the environment) reaction had occurred? Judging by its pH is the HNO_3 an acid, a base, or neutral? Judging by the final pH, is the product mixture an acid or base or neutral? From the mass of the potassium sulfite, what is the number of moles of it at the start of this chemical reaction? From the mass of the HNO_3 how many moles of it are present at the start of this chemical reaction? Calculate the number of moles of it at the start of the reaction from its volume in milliliters and its molarity (moles/liter) of 10 Molar. From the number of moles of the two reactants (the potassium sulfite and the nitric acid), how many moles of potassium nitrate theoretically could result (assuming a 100% yield)?

H. Double Replacement Reactions (Precipitation Reactions) in which Solutions of Two Soluble Salts react to make a Precipitate of an Insoluble Salt:

1. Reactants and Products: Calcium Chloride, Potassium Carbonate, Potassium Chloride, Calcium Carbonate (three soluble salts, and a precipitate) - $CaCl_2$, K_2CO_3, KCl, $CaCO_3$. Chemical reaction equation: $CaCl_2 + K_2CO_3 \rightarrow KCl + CaCO_3$. Balanced Equation: $CaCl_2 + K_2CO_3 \rightarrow 2KCl + CaCO_3$.

CaCl2 + K2CO3 ▢ 2KCl + CaCO3.

1 mole of Calcium Chloride=		110.98 grams
1 Mole of Potassium Carbonate=		138.21 grams

	CaCl2	K2CO3	2KCl(sylvite)	CaCO3(calcite)
Delta G				
Physical State	col, cubic	wh hygr pwdr	wh cr cub	col, rh

Matreials:

Calcium Chloride (a soluble salt) - $CaCl_2$

Potassium Carbonate (a soluble salt) - K_2CO_3

Distilled Water (a molecule) - H_2O

Two 150 milliliter beakers

50 milliliter graduated cylinder

Weighing glassine paper

Digital or triple beam balance

Celsius thermometer

pH meter

Glass rod

Procedure:

Tare a piece of glassine weighing paper on a balance. Using the balance, weigh out 6 grams of calcium chloride on the paper. Label one of the empty (clean, dry) 150ml beakers with the characters $CaCl_2$ and weigh it on the balance. Record this as the empty $CaCl_2$ beaker weight. Transfer the 6 grams of calcium chloride into it and weigh it again. Record this as the beaker weight with $CaCl_2$. Pour 20 milliliters of distilled water into the 50ml graduate cylinder (be sure to measure from the bottom of the meniscus). Decant (pour) the 20ml of distilled water into the 150ml beaker containing the calcium chloride and weigh it again. Record this as the $CaCl_2$ beaker with water weight. Tare a piece of glassine weighing paper on a balance. Using the balance, weigh out 7 grams of potassium carbonate on the paper. Label the other empty (clean, dry) 150ml beakers with the characters K_2CO_3 and weigh it on the balance. Record this as the empty K_2CO_3 beaker weight. Transfer the 7 grams of potassium carbonate into it and weigh it again. Record this as the beaker weight with K_2CO_3. Pour 20 milliliters of distilled water into the 50ml graduate cylinder (be sure to measure from the bottom of the meniscus). Decant (pour) the 20ml of distilled water into it and weigh it again. Record this as the K_2CO_3 beaker weight with water. Measure the temperature of the distilled water in the two beakers with the Celsius thermometer (rinse the thermometer with distilled water between measurements). Record it as the starting temperatures. Stir the contents of these beakers with the glass stirring rod (rinse it between beakers). Be prepared to observe all changes to the calcium chloride solution when you add it to the potassium carbonate solution. Decant (pour) the 20ml of calcium chloride solution into the 20ml potassium carbonate solution. Stir the resulting solution with the glass rod. Visually observe the changes that occur in the resulting solution. Measure the resulting solution's temperature (call it the final temperature). Let the precipitate settle. Decant as much of the solution in this beaker as possible while leaving all of the precipitate in the beaker. Allow the beaker to dry. Weigh this beaker and record it as the $CaCO_3$ beaker mass.

Discussion and Conclusion Questions:

Write the balanced equation of the reaction that you have carried out: $CaCl_2 + K_2CO_3 \rightarrow KCl + CaCO_3$. What type of reaction is this (combination, decomposition, substitution)? Does the temperature change indicate that an exothermic (heat was given off to the environment) or endothermic (heat was absorbed from the environment) reaction had occurred? From the mass of the calcium chloride, what is the number of moles of it being used in this chemical reaction? From the mass of the potassium carbonate, how many moles of it are being used in this chemical reaction? From the number of moles of the two reactants (the calcium chloride and the potassium carbonate), how many moles of calcium carbonate theoretically could result (assuming a 100% yield)? From the mass of the $CaCO_3$ how many moles of it were produced and what is the % yield? What role does the water play in this reaction?

2. Reactants and Products: Silver (I) Nitrate, Iron (III) Chloride, Silver (I) Chloride, Iron (III) Nitrate (two soluble salts, a metal salt, and a precipitate) - $FeCl_3$, $AgNO_3$, $AgCl$, $Fe(NO_3)_3$. chemical reaction equation: $FeCl_3 + AgNO_3 \rightarrow AgCl + Fe(NO_3)_3$. Balanced Equation:

$FeCl_3 + 3AgNO_3 \rightarrow 3AgCl + Fe(NO_3)_3$.

FeCl3 + 3AgNO3 ▢ 3AgCl + Fe(NO3)3.

1 mole of Iron (III) Chloride=	162.2 grams
3 moles of Silver Nitrate=	509.64 grams

	FeCl3(molysite)	3AgNO3	3AgCl(ceragyrite)	Fe(NO3)3.(fe(iii)9 hydrate)
Delta G				
Physical State	blk-brn, hex	col rh	wh cub	lt vlt, mn, deliq

Matreials:

Silver (I) Nitrate (a soluble salt) - $AgNO_3$

Iron (III) Chloride (a soluble salt) - $FeCl_3$

Distilled Water (a molecule) - H_2O

Two 150 milliliter beakers

50 milliliter graduated cylinder

Weighing glassine paper

Digital or triple beam balance

Celsius thermometer

pH meter

Glass rod

Procedure:

Tare a piece of glassine weighing paper on a balance. Using the balance, weigh out 10 grams of silver (I) nitrate on the paper. Label one of the empty (clean, dry) 150ml beakers with the characters AgNO₃ and weigh it on the balance. Record this as the empty AgNO₃ beaker weight. Transfer the 10 grams of silver (I) nitrate into it and weigh it again. Record this as the beaker weight with $AgNO_3$. Pour 20 milliliters of distilled water into the 50ml graduate cylinder (be sure to measure from the bottom of the meniscus). Decant (pour) the 20ml of distilled water into the 150ml beaker containing the silver (I) nitrate and weigh it again. Record this as the $AgNO_3$ beaker with water weight. Tare a piece of glassine weighing paper on a balance. Using the balance, weigh out 3 grams of iron (III) chloride on the paper. Label the other empty (clean, dry) 150ml beakers with the characters $FeCl_3$ and weigh it on the balance. Record this as the empty $FeCl_3$ beaker weight. Transfer the 7 grams of iron (III) chloride into it and weigh it again. Record this as the beaker weight with $FeCl_3$. Pour 20 milliliters of distilled water into the 50ml graduate cylinder (be sure to measure from the bottom of the meniscus). Decant (pour) the 20ml of distilled water into it and weigh it again. Record this as the $FeCl_3$ beaker weight with water. Measure the temperature of the distilled water in the two beakers with the Celsius thermometer (rinse the thermometer with distilled water between measurements). Record it as the starting temperatures. Stir the contents of these beakers with the glass stirring rod (rinse it between beakers). Be prepared to observe all changes to the iron (III) chloride solution when you add it to the silver (I) nitrate solution. Decant (pour) the 20ml of iron (III) chloride solution into the 20ml silver (I) nitrate solution. Stir the resulting solution with the glass rod. Visually observe the changes that occur in the resulting solution. Measure the resulting solution's temperature (call it the final temperature). Let the precipitate settle. Decant as much of the solution in this beaker as possible while leaving all of the precipitate in the beaker. Allow the beaker to dry. Weigh this beaker and record it as the $Fe(NO_3)_3$ beaker mass.

Discussion and Conclusion Questions:

Write the balanced equation of the reaction that you have carried out: $FeCl_3 + AgNO_3 \rightarrow AgCl + Fe(NO_3)_3$. What type of reaction is this (combination, decomposition, substitution)? Does the temperature change indicate that an exothermic (heat was given off to the environment) or endothermic (heat was absorbed from the environment) reaction had occurred? From the mass of the iron (III) chloride, what is the number of moles of it being used in this chemical reaction? From the mass of the silver (I) nitrate, how many moles of it are being used in this chemical reaction? From the number of moles of the two reactants (the iron (III) chloride and the silver (I) nitrate), how many moles of iron (III) nitrate theoretically could result (assuming a 100% yield)? From the mass of the $Fe(NO_3)_3$ how many moles of it were produced and what is the % yield? What role does the water play in this reaction?

Bibliography

Phillip Ball, *The Ingredients: A Guided tour of the Elements*, Oxford University Press, Oxford England, 2002 ISBN 0-19-284100-9

Footnotes

[1] page 23, *The Ingredients: A Guided tour of the Elements*, Phillip Ball, Oxford University Press, 2002

[2] page 29, *ibid*
[3] page 30, *ibid*
[4] page 30, *ibid*
[5] page 30, *ibid*
[6] page 34, *ibid*
[7] page 34-35, *ibid*
[8] page 35, *ibid*
[9] page 38, *ibid*
[10] page 39, *ibid*
[11] page 39, *ibid*
[12] page 40, *ibid*
[13] page 40, *ibid*
[14] page 42, *ibid*
[15] page 87, *ibid*
[16] page 29, *ibid*
[17] page 86, *ibid*
[18] page 86, *ibid*
[19] page 87, *ibid*
[20] page 87, *ibid*
[21] page 102, *ibid*
[22] page 105, *ibid*
[23] page 93, *ibid*

About the Author

Greg Gebhart has a BA in Chemistry from Swarthmore College, is All But A Dissertation in Biochemistry from University of Buffalo, has a MBA in Operations Management from Vanderbilt University, and has a Med in Science Education from University of Houston.

www.ingramcontent.com/pod-product-compliance
Lightning Source LLC
Chambersburg PA
CBHW051201220526
45473CB00003B/862